INTERPLANETARY LIBERTY

INTERPLANETARY
LIBERTY
BUILDING
FREE SOCIETIES
IN THE COSMOS
by
CHARLES
COCKELL

The frontispiece was designed by Charles S. Cockell in collaboration with artist Harry Brockway, who made the print. It is a wood engraving that takes its inspiration from, and is a revision of, the frontispiece of *Leviathan* by Thomas Hobbes, published in 1651.

In the upper part of the image, we see a giant in a spacesuit, symbolising the power and authority that materialises in the extraterrestrial environment from the control of vital resources. Unlike Hobbes's leviathan, it does not emerge by agreement and contract between the people, but ineluctably emanates from a society in extremity (as represented in the body of the giant). The power it wields is represented by a tube providing oxygen, which the giant holds in its right hand, and a hose providing water, in the left hand. In contradistinction to Hobbes's leviathan, because this concentrated extraterrestrial power involuntarily creeps into society, it must be energetically directed and controlled by the people if it is not to drive society into abject tyranny. In the landscape below, around a Mars or lunar station, we see the space settlers pulling ropes attached to the giant to attempt to control and direct its power. The ropes pull on the oxygen and water supplies, since it is through those supplies (and others besides, such as food and power) that the power of authority will manifest.

In the lower part of the image, we see some depictions of the different types of confinement in the extraterrestrial environment that lead to this challenge with power. Clockwise from top right, they are: a lander on another world, a station on the surface of another world, spacesuits required to venture outside, a rocket ship travelling between locations in space, a space station in the free plateau of space or in orbit around a planetary body, and a pressurised vehicle on the surface of a planetary body.

INTERPLANETARY LIBERTY

Building Free Societies in the Cosmos

CHARLES S. COCKELL

School of Physics and Astronomy,
University of Edinburgh, UK

OXFORD
UNIVERSITY PRESS

OXFORD
UNIVERSITY PRESS

Great Clarendon Street, Oxford, OX2 6DP,
United Kingdom

Oxford University Press is a department of the University of Oxford.
It furthers the University's objective of excellence in research, scholarship,
and education by publishing worldwide. Oxford is a registered trade mark of
Oxford University Press in the UK and in certain other countries

Impression: 1

Published in the United States of America by Oxford University Press
198 Madison Avenue, New York, NY 10016, United States of America

British Library Cataloguing in Publication Data
Data available

Library of Congress Control Number: 2022930880

ISBN 978–0–19–286624–0

DOI: 10.1093/oso/9780192866240.001.0001

Printed and bound in the
UK by Clays Ltd, Elcograf S.p.A.

PREFACE

Liberty! . . . let us repeat it without fear of wounding any power that deserves respect; for all that we love, all that we honour is included in it. Nothing but liberty can arouse the soul to the interests of social order.

Germaine de Staël[1]

Alexander Hamilton, one of the framers of the US Constitution, asked a question that has a universal applicability—and when I say universal, I mean it literally. He pondered, in the first of *The Federalist Papers* 'whether societies of men are really capable or not of establishing good government from reflection and choice, or whether they are forever destined to depend for their political constitutions on accident and force'.

Hamilton and his fellow framers wanted to know whether humans were capable of breaking free from the age-old habit of constructing societies wherein coercion, mixed with the unpredictabilities of history, could be replaced by some sort of rational order that would be developed, modified, improved, and optimised by human choice. Could we move beyond the wretched state of human despotism that characterised so much of the ancient world, apart from flickers of liberty in places such as ancient Athens, to a better world where the liberty and happiness of humankind could be secured by choice? The question he asked was the prelude to the construction of an American constitution which would attempt to answer it in the affirmative and suggest the means by which it was to be done.

Yet the question Hamilton asked has no tincture of US-centrism; it was not a question born of the eighteenth-century atmosphere of revolution in which the framers found themselves. It is a universal question that we could ask about any group of human beings who find themselves in a state where they must construct a new society. Are they to submit to the force of a tyrant and the unwieldy accidents of human society or nature, or are they to attempt to fabricate for themselves something over which they have a measure of control?

In simple terms, the purpose of this work is to address this question for the environments beyond Earth, for the vast spaces and innumerable worlds that we might explore and settle in the decades, centuries, and millennia to come. Implicitly throughout, I am considering settlements on the surface of other planetary bodies such as the Moon and Mars, in the open plateau of space in our Solar System, and potentially world ships heading across the vastness of interstellar space towards distant stars. It is human communities that concern me. I will have less to say about the applications of liberty and legality to matters such as satellite ownership, the environmental impact of launch sites, and many other domains in which questions of freedom intersect with space exploration; I am primarily concerned with the political-philosophical aspects of space with respect to human organisation.

I have two fundamental propositions I wish to advance in this work, which will run as guide rails throughout. The first is that space is tyranny-prone. Its extremities and lethality do not assure the emergence of despotic regimes, but they encourage them. Space is a dictator-friendly place. I seek to lay out that landscape, its contours and valleys, and to point out to the reader the dangerous canyons into which a society might fall, mapping, if you will, the hazards to which extraterrestrial settlers must forever be vigilant as they embark on their social expedition into the universe. My second proposition is that we can veer humanity away from these precarious obstacles. Thought, political culture, education, and even engineering can be used to construct free societies in the cosmos. Much of this book is spent discussing some means by which we might realise these ends. My purpose, if you will, is to suggest the design of bridges and overpasses by which society might cross the ravines that I have drawn attention to or even avoid them entirely.

To take the lazy route, let me make a small change to Hamilton's query, shorn of its eighteenth-century bias. The question that I will address is 'whether extraterrestrial societies are really capable or not of establishing good government from reflection and choice, or whether they are forever destined to depend for their political constitutions on accident and force'. Determining the types of societies that we construct in space and the extent of the liberties that they might achieve is, for me, one of the most fascinating and extraordinary challenges of political philosophy in the twenty-first century and beyond.

I want to point out what this work is not. It is not a complete account of the characteristics of an extraterrestrial society, or a prediction of what it might be. There are many wonderful, and very elaborate, science-fiction depictions of future societies in our Solar System and further afield which lay out the details of social arrangements, eating provisions, office arrangements, and countless other minutiae; you will find few of these details discussed in this work. My singular focus is on the problems of liberty at the large scale, and although one could, with some conviction, argue that this matter will weave its way into almost every facet of society, my aim is not to pursue exhaustively every recondite avenue in which it might be expressed or to describe the full range of the possible consequences. I will not deliberate the matter of, say, office or eating arrangements, although it is clear that the political nature of society could well influence such designs. I will not discuss the details of laboratories or similar facilities, although I draw attention to the importance of science. Rather, I will discuss what I think are the main social elements influenced by the conditions for liberty in space and their repercussions for the wider political, cultural, and economic environment. Another reason for this broad-brush approach is that there is the challenge that in everyday experiences, it is very difficult to predict what an extraterrestrial society will be like; this is in essence the problem of the resolution and scale of credible prediction. Even at the levels I grapple with here, I make no claims that my views will turn out correctly. However, I do think it is possible to navigate a discussion on freedoms in space, given what we know about the physical nature of that environment, without discussing the finer matters of life. Nevertheless, in some places in this work, such as on the subject of education, I will make some specific suggestions and observations.

The reader might get the impression, as they make their way through this book, that the negative prognosis that I give for liberty in space suggests that I am similarly negative about the human prospects beyond Earth; this is indefatigably not the case. I am an enthusiastic proponent of the human future in space and I regard our space-faring, eventually star-faring, capacities and opportunities as nothing short of a remarkable possibility that we would have to have the most astonishing myopia not to seize and instead allow ourselves to join the approximately 99 per cent of species that have gone extinct on Earth. My jaded view of the prospects for freedom beyond Earth may well turn out to be too negative (I hope it does), but in assuming some of the worst-case scenarios and directions of autocracy in space, we are better placed to mitigate them by intentionally building the institutional structures that can steer us away from them. My view is that if you want to build good societies, it is better to aim for anti-dystopias than to try to construct never-achievable utopias. I apply this general view to my analysis of the space frontier.

I think it is necessary for me to be clear to the reader about my own hue of political philosophy. No matter how objective we try to be in these matters, the way in which we view human institutions, and more so the way in which we suggest how they should be improved, is fashioned by how we think humans behave individually and collectively. These questions are strongly value-laden and influenced powerfully by our personal views. To attempt to pass this work off as an effort in complete objectivity in considering the types of institutional arrangements we might want to create beyond Earth would, I think, be disingenuous. There is one facet of my own view on political philosophy that the reader should know about, which has an inescapable influence on the conclusions I draw throughout; I leave the reader to decide whether this view fatally biases my conjectures or renders my suggestions compromised, but I wish to make it clear anyway.

I align strongly with a classical liberal view of how I think human societies are best managed. By this I mean that although I have always attempted to read ideas across the political spectrum, I find myself most sympathetic to the generality of ideas, if not always the specifics, of writers such as Locke, Montesquieu, Mill, Smith, Madison, Hayek, and others. The reason for this predisposition is not that I have a blinkered affiliation to the ideas of Western white male political philosophers. The last 10,000 years of settled human history have brought forward a vast and irrepressible series of experiments in the fabric of human government. In ancient times, the military society of Sparta lay side by side with the direct democracy of ancient Athens. In more modern times, we witnessed the democratic republic of the United States of America eyeballing the communist-inspired centralised state of the Soviet Union. Almost every brand of society, from individualist to highly collectivist, in almost every geographical region, and across the expanding size of the human population, from village committee to continental-scale federation, has been attempted. No doubt we have not exhausted the potential landscape of political experiments, but we have already achieved an impressive variety of them.

All things considered, I think that systems of government in which individuals charged with governance consider themselves to be the servants of the people, acting to implement the choices of the populace, have brought forward much greater felicity in societies over the long term than structures where the government sees itself as the autocratic wielder of power, the principal director of human activity. With this general view in mind, the ideals of accountable government, voluntary association, the rule of law, equality before the law, and respect for the many and varied individual visions of the good life are those that recommend themselves for the construction of governments. I believe that these forms of power are more likely not just to yield happy citizens, but also to inculcate into individuals and society the essential virtues of self-governance.

I have attempted where possible to avoid starting with my own views, but rather to lay out the facts about the extraterrestrial environment as they are and then to deduce what sorts of institutional structures might lend themselves to dealing with these realities in a logical way. It so happens, for reasons I will elaborate, that I end up concluding that many ways to reduce the despotism inherent in the space environment include adopting and modifying classical liberal approaches. In other words, I have ended up arriving at conclusions that are congruent with a view of political philosophy that I already hold. The reader may find this rather convenient, suspicious even. That is precisely why I declare my own view of society from the outset, so that the reader may be alerted and may make their own judgement on the conclusions I draw.

I wish to say something about the motivation for this book. I am not a political philosopher by training. I obtained my degree in biochemistry and my doctorate in molecular biophysics. My interest in political philosophy was ignited by a chance encounter with Rousseau's *The Social Contract* in the summer of 2000 in a second-hand bookshop in Cambridge, UK. It opened me up to an effusive fascination with political philosophy that naturally spilled into my professional interests in space exploration. I was surprised at the lack of academic treatment of liberty in space, despite the long and rich history in the political philosophy of freedom, and so in 2008 I published a paper on this subject, 'An Essay on Extraterrestrial Liberty', in the *Journal of the British Interplanetary Society*, the journal of that venerable space advocacy group. Since that time, I have, on and off, written other papers discussing different aspects and angles of ideas on how freedoms will be expressed in the extremities of space. This book synthesises a set of some of my thoughts on the subject of freedom beyond Earth.

Some people might feel uncomfortable with my cavalier treatment of professional boundaries in social sciences by launching into papers and books without any prior formal philosophical training. I humbly demure to that opinion, but I am much more convinced and impressed by the general argument that *all* human beings should engage in vigorous and energised debate about how their societies should be run, and none of us should feel reticence about expressing views or speculating on how to build good societies. Such is the importance of political philosophy to the human condition, that it is one subject area that should not admit of academic limitations; political philosophy should be a subject of feverish interest and discussion among all human beings in restaurants, homes, and workplaces, on Earth and in space. Thus, my conceit in throwing myself wholeheartedly into writing political philosophy is a reflection of how important I think this field is.

Despite these comments, my professional interests have usefully intersected with this task. I have spent my career involved in space science and thus part of the equation—an understanding of the physical conditions around which the institutions of extraterrestrial government must be constructed—is familiar to me. I spent 10 years working for NASA and the British Antarctic Survey. During this time, many field seasons working in Antarctica, the Canadian High Arctic, and other deserts have provided experience from which I draw some intuition for the observations I make about the effects of extreme environments on social organisation. More recently, I have spent eight years informally studying the social history and political arrangements of the small Scottish Hebridean islands from the point of view of extraterrestrial settlement, analysis I draw upon in this work.

There is another rationale for this interest. The environmental conditions in space have profound effects on biology. Across the pantheon of life, everything from microorganisms to rats have been investigated in the extremes of outer space. Attempts to understand the underlying mechanisms of how radiation and gravity, for instance, influence organisms is an intense field of investigation. Since human beings build societies, not only are we interested in comprehending how space affects the biochemistry and physiology of the human organism, but we might also take an interest in understanding how extraterrestrial extremities shape the communities that we create. Therefore, in some sense, one could regard political philosophy in space as a continuation of that biological interest, extending from the organism into the more intractable questions of human behaviour and social organisation. For me, as a biologist, the problem of political philosophy beyond Earth is a natural and fascinating extension of my own interest and work investigating biological systems in space.

Finally, I wish to point out that my conclusions here are subject to considerable uncertainty. I won't shy away from an explicit and forceful statement that this is a work of speculative political philosophy. It is a future projection, not a disquisition on historical matters, which is the more usual form of political philosophy. Without the facts and circumstances of the past to provide tram lines to guide and steer hypotheses, it is necessarily infused with views and conjectures that are yet to be tested. We have not constructed societies in space, and the way in which they emerge may throw up surprises in human organisation that will flummox us. Yet, we might also observe that this intellectual terra incognita beckons us towards a rich and interesting future of political philosophy in space.

I do not claim that my analysis touches on all aspects of the liberty of human settlers beyond Earth. It is just one contribution, one particular view, one treatise in a canon of thought that might be directed towards this problem. Despite that, my discussion rests on the assumption that there are, in fact, characteristics that in the absence of the engineering of the human brain itself (a matter I will address) will remain the same in space as on Earth. These essential universals, if you will, about human behaviour and our societies allow us, I believe, to fruitfully contemplate the future of human societies beyond Earth and to advance the prospects for free societies and worlds in the cosmos.

Charles S. Cockell
Edinburgh, January 2021

ACKNOWLEDGEMENTS

I would like to thank the many colleagues and friends with whom I have had conversations on the exploration and settlement of distant worlds and the forms of freedom that may be possible there. It is invidious to name them as it invites accidental omission, but many of them are to be found in the citation list of this work, associated with the multi-authored books we wrote following the three meetings we held at the British Interplanetary Society on the subject of 'Extraterrestrial Liberty' from 2014 to 2016. Preceding those meetings, my career had been scattered with conversations with other colleagues and friends, particularly at the NASA Ames Research Center and the British Antarctic Survey, where some of the thoughts expressed herein took root. Somewhat surreal conversations on extraterrestrial governance have blossomed over tea or coffee on Antarctic glaciers and in a polar asteroid crater. I would like to thank those friends for their insights. These discussions, embedded as they were in environmental extremes, had a form of authenticity and force that gave shape to some of the discussion here. I also thank the islanders of the Hebrides who over the past few years have welcomed discussion of their island governance structures with the intention of ascertaining how to live in space.

Finally, perhaps unconventionally, I would like to project some acknowledgements into the past and future. Although they are not here to appreciate it, I would like to acknowledge the astonishing writing of those who sparkled with the thought of free societies. Perhaps because of the revolutionary events in America and France, and the achievement of the full energy of the Enlightenment, the writers of the eighteenth century have particularly inspired me with their passionate views, their desire to see freedom across human societies blossom, and the sheer vivacity and quality of their writing. However, that energy and assertiveness also fills the pages of thoughts on liberty from ancient Athens and Rome. Perhaps because it involves human institutions on which the quality of lives depend, generally I find political philosophy literature some of the most remarkable and arresting writing. I have appreciated the privilege of reading it.

Finally, I would like to project an acknowledgement into the future, to those living permanently in space. Freedom in societies beyond Earth has driven much of my motivation for reading political philosophy and writing this book. I do not know if any of what is written in this book will have relevance to the situation that ultimately emerges. However, I hope that at a minimum it will contribute to a general keenness and alertness to advance liberty into the cosmos and to defend the ideas of free human minds and action. To them I say: I wish you well in the construction of free societies. It is a task that I think is of the greatest consequence for the human future.

CONTENTS

1

LIBERTY ON THE SPACE FRONTIER

Liberty: the endless human challenge on Earth and in space

Thomas Hobbes, a seventeenth-century political philosopher, was not optimistic about humanity's chances of successfully building societies that would not descend into incessant war, each individual pitted against each other in a nasty, brutish, and short life, without some sovereign to watch over them—and a Leviathan of a sovereign, no less.[1] He was not the first to hold this view and he certainly was not the last. The origin of Hobbes's sentiment is easy to grasp in the modern world. It is none too difficult to watch the news and, if one is a peacefully minded person, develop a sense that people need to be led, ordered, and organised to ensure felicity and tranquillity in society. Yet his view was not universal, and it was soon countered by John Locke[2] and others, who viewed the ability to overthrow a sovereign or government that had lost the approbation of the public as not merely desirable but a fundamental right. The members of a society were not to be cowed and controlled by an all-imposing manager but instead were to engage in a more restricted social contract to construct a polity where some aspect of their freedom was to be given up to secure their property and their lives, while allowing for a wide range of personal freedoms.

And so in this period, known as the Enlightenment, an argument raged. This was not a new argument, but an existing one given new impetus—and it raised the eternal questions of any settled human society: what level of control and freedom is appropriate and right for society? How can we balance the need for security with the boundaries of individual freedom? As Thomas Hobbes put it: 'For in a way beset with those that contend on one side for too great Liberty, and on the other side for too much Authority, 'tis hard to passe between the points of both unwounded.'[3] Hobbes could see that it was hard to get the balance right.[4] One might even be as bold as to suggest that the human disagreement about the modes of societal organisation that are necessary to get this balance right has been the dynamo at the heart of almost all systems of human governance proposed. In contemplating the history and future of India, Nehru[5] had this to say about the equilibrium between liberty and authority: 'That conflict is not of India only; it is of the West and also of the entire world, though it takes different forms there.'

Of course, as political philosophers have realised, the balance between these two extremes can never be reached with dogmatic certainty.[6] Indeed, surely it is the mark of a free society that the point between too much authority and too much liberty is constantly under revision as society, technology, and the wider world change; as soon as a society settles the fulcrum in one place, new innovations, social situations, and mores are likely to add or take away weights from liberty or authority, causing the balance to shift. Hobbes was right: it is difficult to pass between liberty and authority without being wounded. To this we might add that the perceptions of where that balance lies vary greatly between individuals. Even if a ruling authority was to judiciously and successfully divine where between the two extremes the balance lies, there would be some distribution of humans on either side of this point who would perceive too much authority or too much liberty. In other words, the balance can never be right because even if it is agreed to be well judged as an average, individual proclivities and morals will never admit of a state that makes all members of a society happy.

We might define a free society not as one that has successfully achieved some abstract idea of this balance but instead as one that is open to a continuous and never-ending re-evaluation of this trade-off as conditions change; a free society is sensitive to the altering ideas of its members and it works to modify its institutions and structures of governance to best accommodate these shifting sands.

This challenge of navigating between authority and liberty has been traced through human society ever since the first civilisations. In an unconscious way, it no doubt precedes even these settled communities in the form of dominance displays and hierarchies in primate populations.

Now a new vista on this perplexing problem opens up: the expansion of humanity into space.[7] In the 1950s, after the first satellite was launched into space, the matter of freedom beyond the confines of Earth came into full view. It is true that freedom in space did not impinge directly on the lives of a great many people, since there were no humans permanently living in space, but the space age did, nonetheless, almost immediately bring forward questions of liberty.

The activities of humans outside the limits of Earth raised immediate questions about sovereignty in space. Fearful that a rush to touch down on planetary bodies such as the Moon would ignite highly polarising and nationalistic claims on that land, the UN Outer Space Treaty and the Moon Treaty essentially forestalled national claims and, in the latter case, left the Moon and other bodies as the 'common heritage of mankind'; but within this determination was a statement about the freedom, or not, of people to take ownership of lands beyond Earth. Whatever our view on these provisions, it was an all-encompassing attempt to circumscribe, or at least to define, what liberties humans might have after leaving Earth and an illustration that we cannot, by simply leaving our planet, escape the eternal argument that Hobbes so clearly identified.

Then there was the matter of whether we might contaminate other planetary bodies with microorganisms, or even return alien life to Earth and compromise our biosphere with these unintentionally maleficent creatures. These concerns did not enter into international law, but they were eventually given form by the Committee on Space Research (COSPAR), which brought into being a set of recommendations on the cleanliness of spacecraft to prevent the harmful contamination of other planetary bodies.[8] Craft visiting planets where we might seek alien life were considered to threaten or compromise our search for that life by introducing the terrestrial life carried on them; this was called 'forward contamination'. Likewise, those craft bringing samples back from these locations might release a life form that endangered our own biosphere: 'back contamination'. Planetary bodies with no chance of an origin of life or an indigenous biosphere placed little constraint on spacecraft designers and builders, but places like Mars, where the presence of indigenous extant life was an open question, would be subject to these recommendations just as they are today.

These decisions on spacecraft cleanliness impacted the freedom of spacecraft designers, who now had to consider how to design a craft that could be satisfactorily rendered cleansed of microbes and other organic material to some defined standard. The budget setters now had to factor the costs into their expenditure to make those craft pristine. And here again, no matter what a person's views on the specifics of these policies, an ineluctable link between space exploration and human liberty was forged.

Thus, although it may seem that the concern for freedom and its manifestations beyond Earth is a new one, as soon as humans ventured into the heavens, even with robots, it was impossible for them to escape the problems of liberty. Property rights and microorganisms, to name just two complications, rudely forced freedom onto the agenda. To these two we might add early discussions about the freedom to use certain satellite orbits or the laws that would govern who was to be responsible for space debris returning from orbit and unintentionally landing in someone's house. Hobbes and Locke became interplanetary long before anyone deliberately considered what these philosophers might have had to say about humanity's movement into space.

The legalities and rules that surround the robotic exploration of space and their consequences for liberty continue to be, and will always be, of importance to nation states and private entities and how they operate in space. However, an especially interesting set of challenges is opened up by the human exploration of space and, in particular, the prospect of permanent human settlements. Once we have collections of settled humans in space, we can see at once a fusion between knowledge of the new and extreme environments that they will inhabit and the questions of political philosophy that have absorbed the human imagination for millennia. Those age-old questions—say, 'Can humans be free?' or 'What do we mean by freedom?'—find new expression in the vastness of the Solar System and the spaces beyond.

We might ask ourselves: why would these questions be any different from the situation on Earth? We could consider Mars, for example, to be just another place. Just as the conditions for liberty in Asia do not necessarily open up completely disconnected questions in political philosophy compared to Europe, say, but rather involve discussions around similar, and sometimes very old, questions,[9] so too surely the expansion of humans beyond Earth extends the spatial realms of a set of questions that remain largely undisturbed.

In some sense, we would be right to consider space to be just another place, a separate location in which to apply ancient questions about the extent and nature of freedom. Many of the questions I discuss in this work deviate little from questions asked of societies on Earth.[10] Yet what makes the space frontier so fascinating in this regard is not the questions themselves, but the sheer extremity of the conditions that influence the answers we give to questions about liberty and the way in which freedoms might be manifested in these environments. This series of conjectures and thoughts is given extra vitality by the fact that at the time of writing, we have not established any permanent self-sustaining settlement in space, so we know little about how those societies will emerge, making the nature of freedom beyond Earth a matter still open to much spirited and vigorous debate.

It is possible to deliberate on the problem of liberty for future space societies

It would be apropos for a person to point out that the lack of permanent settlements in space (other than the semi-permanent

International Space Station) would make speculations on the nature of liberty beyond Earth rather weak; perhaps we should wait until we actually have a society with which to theorise, as almost every political philosopher who has cogitated on the nature of human organisation on Earth has had. However, the assumption that underpins this work is that the human personality remains largely invariant and the same emotions and basic characteristics that shape human societies on Earth will express themselves in space. Therefore, when we think about freedom in space, we are not dealing with an entirely blank canvas, but with material that behaves in certain ways and which has been observed already for millennia, albeit under terrestrial conditions. As Christakis simply observed: 'But human societies do not come from somewhere else. They come from within us.'[11]

Again, the reader might be inclined to take up a powerful rejoinder: societies on Earth have taken on a myriad of forms and shapes that differ markedly and diverge in infinite ways in their subtleties. Indeed, the field of anthropology would not exist if this was not the case. We could wonder, then, whether any discussion on liberty beyond Earth would not be seriously flawed by the innumerable ways in which societies might develop beyond Earth. I think these anxieties are quite valid. To attempt to describe the exact structure of a society in space and all of its peculiar characteristics and mores would be pointless (although perfectly entertaining, hence the whole genre of space science fiction). To attempt to lay out the expected characteristics of an extraterrestrial society in all its details as if this were an academically rigorous description of a future society, and how it *will* be, would be futile.

The tract I embark on here is therefore not an attempt to engage in that usually inaccurate activity of predicting the future in meticulous detail.[12] My purpose is rather to answer the question of how the conditions in space will generally alter the landscape for different types of human liberties and from that position to make some suggestions for how forms of despotism that may be antithetical to human happiness could be deliberately thwarted. One might say that this work attempts to get at some broad contours and outlines of human liberty without attempting to define or predict how exactly human societies will emerge.[13]

My contention that we can define certain characteristics of how humans will operate beyond Earth and how species of freedoms that have been familiar to us throughout our history will manifest beyond this planet is, of course, an assumption. I have absolutely no hesitation in believing that if one were to read this book in the far future, the arguments would seem archaic, and in places downright wrong; but I would hope that in their general gist they might be right. Of course, if I did not think that, I would not write it, but I defer to the future the possibility that even this supposition may be wrong.

There is an observation that I think is relevant here. When we sit down to read the lives of famous Romans or Greeks penned by Plutarch, or we consider Polybius' or Cicero's thoughts on the design of a republic, we do not find ourselves dumbfounded and perplexed by what we read, as if we had been granted readership of a manuscript written by aliens from a distant planet whose minds and society were of a completely different architecture. Instead, we find that their observations comport with our own perceptions about human conduct and society. When we read Epictetus' advice on how to deal with the powerful or the inconsistent, we are not confused by his insights. Rather, we understand them completely, and we may even find ourselves, when we read his advice on the adulterous or on people who are prone to lightly share their personal information, brought to laughter, as no doubt readers were more than a thousand years ago. Had I been born at any given point in the last two millennia,

I could have knowingly nodded to Epictetus about the goings-on of my neighbours or the rapacity of some politicians as if we were sharing these views over a coffee.

Our societies, technology, capacity for science, means of transport and communication, ways of cooking and eating, medical knowledge, and so forth have advanced so as to be unrecognisable to Epictetus, but our basic behaviour has not changed. It is this constancy in the form of human interactions, when all around them are in flux, that allows us to say something about the likely hue and characteristic of future human societies, even when we know that those citizens will inhabit a scientific and technological environment that would seem almost magical to us. We could furthermore observe that this constancy across time has been little affected by the occupations of individuals. The types of professions on offer to the Roman populace were very different to many of those that make up society today. In the same way, although extraterrestrial societies may be inhabited by a collection of people that range from scientists and miners to tourists, adventure seekers, and many others, I contend that we do not need to know the human material from which such a society is moulded to believe that the human condition will be much the same. From this claim, I believe that there follows nothing naïve in the idea that we might speculate with some value on the future of liberty beyond Earth. One exception I will admit is that if we should alter the human brain itself, then we might effect a change in the fundamental source of our conduct, and I will have something to say about that later.

As with the ancients, however, what we mean by freedom or liberty is subject to interpretation and context.[14] We could talk about the metaphysical idea of whether we have 'free will'; we could discuss the liberty of the individual to carry out their own ends; we might discuss the freedom of nations to embark on certain courses of action; and we could talk about the freedom of the whole of humanity to act in certain ways, for instance in the face of environmental change. There is no single definition of the words freedom and liberty, although people are wont to exclaim these words as if there were.

In this work, I am primarily occupied with liberty in terms of the capacity for individual human beings, or collections of them, to secure a domain of private thought and action, to think and act freely on their own or with other individuals and groups they choose to, and to live free of undue coercion, domination, or interference from others in the extraterrestrial environment.[15] As with any discussion on these forms of liberty, there can be no quantitative statement to define what we mean. What, exactly, is 'undue coercion'? The point is the pursuit of a culture that respects the choices of individuals and groups and seeks to give those choices a field of action.[16] An essential part of that culture is the capacity for individuals and groups to take part in a free and open discussion about what they mean by liberty in the first place and what they seek to achieve in advancing it. We might say that freedom is the capacity to engage in unhindered private and public argument about the objectives of freedom itself and to be able to fashion a society that promotes the implementation of the results of those deliberations.

Let me also underscore that the realisation of such a state of affairs requires individuals to express the virtue of respecting and encouraging these liberties in others. Freedom is not a crass demand for self-realisation, but a mix of that self-actualisation with the responsibilities and duties towards others to support the wider social structure in which these liberties are not degraded into a set of demands placed against others and society as a whole. Frankl observed:[17]

> Freedom, however, is not the last word. Freedom is only part of the story and half of the truth. Freedom is but the negative aspect of the whole phenomenon whose positive aspect is responsibleness. In fact, freedom is in danger of degenerating into mere arbitrariness unless it is lived in terms of responsibleness. That is why I recommend the Statue of Liberty on the East Coast be supplemented by a Statue of Responsibility on the West Coast.

Although I essentially agree with Frankl's point (and I think his suggestion for a West Coast statue is excellent), responsibility and duty are an inseparable part of freedom itself, not a corollary or the obverse of the same coin. Liberty cannot properly exist at all, or be sustained, in a society unless individuals possess a sense of responsibility and duty to protect the freedom of others, and unless they cherish a willingness to listen to views, and entertain projects, that do not comport with their own outlook. When freedom merely becomes an individual's expectation of rights and non-interference without the corresponding obligations to help build the society that makes those demands possible, then it is not freedom properly so-called; it is nothing more than anarchic selfishness. The recognition of this aspect of liberty becomes of singular importance when we consider the collective efforts needed to build a society in the extraterrestrial environment.

Liberty is my central concern, and the reasons for the obsessiveness over this particular matter beyond Earth will become apparent. However, in maintaining this as my lodestar throughout this discourse, I will discuss how this problem of liberty impinges on governance structures and the whole constellation of human activities, including science, engineering, art, and religion. What are those characteristics of space that I think allow us to discern the major girders and structures that will define the future of liberty on this frontier?

The essential challenge of liberty in space is related to the extremity of the environments to be found there

There are few places on Earth, apart, perhaps, from some regions that circle the equator, where humans could in theory live without any technological support. In most regions of Earth, we, as fundamentally tropical animals, require clothes to survive some portion of the winter, and in many places we require technology, even if only a spear, to collect large items of food. Once settled society emerged, this need for technological artefacts expanded and extended its reach into most items of food, and eventually, as we witness today, even water supplies were managed and canalised to supply the thirst of the many millions of human individuals that were amassed in single locations on the planet.

This pervasive dependency on technology may give us the impression that in space, where environmental extremes also mandate substantial technological support, our questions about human society will be much the same as they are on Earth. We would ask questions about the legitimacy of the state, the way in which people tread the fine line between authority and liberty when living in societies where there is a great deal of mutual interdependence, and so forth.

Whether the social questions we have in space are different as a matter of degree from those we have on Earth, or are something categorically different, I happen to think is not too crucial for the matters that I want to raise in this work; I suspect they are both, depending on the question that is raised. Some aspects of the space environment, such as the amount of food that must be produced and who should control these decisions, have their parallels on Earth; some questions, such as the freedom people have to return to Earth, are distinctly linked to the space environment.

Here, I want to draw attention to the features of the space environment that will define the landscape in which questions of authority and liberty will play out. I will claim from the outset that, whether in degree or categorically, these features define a new vista on these questions that has never been confronted to the same extent on Earth, even in the most extreme environments that have been explored and eventually settled by people. The stark changes in the way in which these questions are manifested merits the examination of space as not just another location for society in which traditional political philosophy and its questions will play a role, but rather as a hitherto unknown frontier in our long adventures with liberty.

Let us focus on what I believe to be the most notable difference, at least with respect to the matters of liberty that we will discuss, between life on Earth and habitation beyond its protective atmosphere: the lack of oxygen to breathe. Oxygen is an essential ingredient in respiration. To place it in its technical context, it is the electron acceptor that carries off all those spent, left-over electrons that provided your body with energy, themselves having come from your sandwiches or whatever else you last ate. As your body is burning these organics all the time to feed its 25-watt brain (with another 75 watts powering your organs, limb movements, etc.), that 100-watt power demand (which fluctuates depending on how much work you are doing) must be constantly supplied. So although you eat only two or three times a day, the products of that food are being continuously processed; those left-over electrons must be removed, hence the need to breathe largely uninterrupted.

Now this brief biochemical diversion may seem a little irrelevant, but the facts of this matter bear directly on the matter of liberty. If it were the case that the human body needed energy only when we ate, then, as with our eating habitats, our requirement for oxygen could be met by infrequent bursts of air intake to match the sandwich eating; but the perpetual need to drive the human machine with its power demands means that although the food can be eaten in spasms (because it can be stored in the body as fats and sugars), the oxygen must be provided continuously. Unlike every other requirement of the human body, it admits of no respite or prolonged disturbance because we do not have the capacity to store it for any appreciable time.[18] Stop breathing and those energy-drained electrons have nowhere to go, and your energy pathways suffer an electron traffic jam, gum up, and stop; in very short order, your body and your brain are deprived of energy and you will die. The time is quick—minutes—maybe a little longer if you are kept cold, but that hardly dispels the essential reality that oxygen is a commodity required in constant supply.

On Earth, that oxygen is provided by the atmosphere, and it is produced by photosynthesis, which spews forth the gas, achieving that prodigious 21 per cent concentration. You might ask how animals were able to exist before oxygen levels were so high in the atmosphere. The straightforward answer is that we think that they could not. The increase in oxygen in the atmosphere seemed necessary to allow for the energy demands of complex multicellular life.[19] Therefore, not only is oxygen essential for complicated creatures like us to persist, but it may even have been the throttle that controlled the timing of the very emergence of animal life.

The pervasive carapace of the atmosphere makes oxygen available to us all.[20] We are denied this gas only when special conditions exist; if you venture into a cave where the oxygen is depleted, your life will be in danger; if you put a bag over your head, the same threat applies. It is very difficult for an authority or a government to impose conditions that deny oxygen, as they do not have the technical means to control the atmosphere on a large scale; however, needless to say, they can, with effort, deny you a breathable atmosphere. Execution by gas chamber as a criminal punishment or the implementation of a programme of genocide, such as the gassings of the Holocaust, are chilling examples of governments essentially taking control of the atmospheric composition. In these cases, they have not so much removed the oxygen as added a poison; nevertheless, they are examples of authorities controlling the gases that people breathe, to their detriment. Conversely, we might also note the myriad ways in which governments and supranational entities have worked to improve air quality; again, this is not so much an attempt to regulate oxygen, per se, as an effort to reduce the background toxicity of other components of the atmosphere. The salient observation is that although governments do have the capacity to change the way in which the atmosphere is used by people as well as its quality, they have very little power to use the quantity of oxygen in the atmosphere as a lever of control over populations and as an easily commandeered method of coercion.

In every extraterrestrial environment beyond Earth, there is either no atmosphere or an atmosphere that cannot be breathed by human beings. Although the search for exoplanets that harbour a biosphere in an oxygen-rich atmosphere is an ongoing activity of astronomy and astrobiology,[21] at the moment we know of no other planet in the universe on which humans might land and freely breathe the atmosphere. Even if we do eventually find such planets orbiting distant stars, and an extraordinary finding that will be, we know empirically that no such planets exist in our own Solar System, the domain within which our human exploration and settlement endeavours will be limited for some time to come. Therefore, for the foreseeable future, whenever humans move beyond Earth, there is an unavoidable connection between technology and a single breath of air that makes the elementary act of breathing a fundamentally technological endeavour. I contend that this single difference lies at the heart of the problem of liberty beyond Earth: the required human intervention to produce something on which we depend upon continuously and which, on Earth, has never been able to be subverted by the heavy hand of government.[22]

Throughout this work, when I discuss this problem, I will talk about oxygen, but I should point out that I am not necessarily implying a pure-oxygen atmosphere; oxygen may be mixed with other gases, including nitrogen and helium, before being used by people. Instead, we could talk more generally about the need for 'air'. However, these other constituent gases are inert carrier gases, and oxygen remains the key physiological molecule required by the human body, so I will give this gas the focus I think it deserves.[23]

I do not contend that this observation concerning the centrality of oxygen in space and its control is new. There is plenty of science fiction—including the now almost cult-status *Total Recall*, in which Martian denizens find themselves at the tyrannical mercy of someone controlling their oxygen—in which these limitations find expression. Nevertheless, it is worthwhile to home in more carefully, with purposeful intent, on exactly why this is a special difficulty and to understand why it is a particularly potent ratchet of illiberality in extraterrestrial environments. More importantly, what can be done to deflect it?

This technical dependence and the consequences for the interdependence it causes between people is given further force by the paucity of other requirements for human existence, particularly water and food. These provisions are not required in a continuous way, akin to oxygen, but they are required over periods of days if thirst and starvation are not to cause widespread social disorder. This temporal buffer, if you like, gives people some leeway to organise dissent against an authority that might control them, but even for the provision of water and food, the difficulties are far removed from the situation in most locations on Earth.

Although there are deserts on Earth where water and food are hard to come by in the wild state, it is ubiquitously the case that beyond Earth there is a lack or extreme deficit of liquid water and there is no food at all that can be harvested from the natural environment. Thus, although the temporal barrier removes the immediacy of the control that others might have, it is nonetheless the case that considerable efforts would have to be made by an individual or small group of individuals to commandeer a fresh and reliable supply of water and food in any location beyond Earth that is not linked to a substantial technical infrastructure. These structures are likely run by people other than, or at least in addition to, those who are seeking these commodities.

These vital victuals can, however, be obtained with work. If they were so denuded from locations beyond Earth, then we could hardly contemplate space settlement at all, or at least we would have to relegate ourselves to a future of a few scattered complexes across the Solar System utterly dependent on resupply from Earth. It is worth exploring briefly some of the technical realities behind the observations I have made to put flesh on these bones.

An energetic and kinetically imposed barrier to extraterrestrial liberty results from the ubiquity of extremes

Let us explore some of the specific environments beyond Earth where humans aspire to settle and the physical challenges in obtaining vital supplies in those locations. On Mars, we have the advantages of the most Earth-like planet in the Solar System; it has abundant reserves of water and an atmosphere that provides some radiation protection. The surface atmospheric pressure is at the triple point,[24] such that large bodies of permanent liquid water cannot be sustained. However, the large inventories of ice, some of it exposed at the surface in the polar regions and near to the surface at more southern latitudes, in principle provides readily available water; it must just be collected, melted, and filtered. Alternatively, it might be retrieved from the atmosphere by condensation.

Once Martian water is collected, oxygen might be obtained by electrolysis of the water, whereby electricity is used to split the water molecule into its constituent hydrogen and oxygen atoms. The oxygen can then be provided for us to breathe, and the hydrogen can perhaps be used to make methane fuel. Oxygen can also be obtained by the 'cracking' of the carbon dioxide atmosphere. In this process, a catalyst, zirconium-based, is used to break up the carbon dioxide molecules to release the contained oxygen atoms. Alternatively, or additionally, oxygen might be produced from plants and photosynthetic algae and cyanobacteria in life-support systems,[25] although these organisms themselves require water.

The surface of Mars, like the Moon, is made of volcanic rock; it is largely basalt. Although it may contain salts which could be toxic to plants and microorganisms at high concentrations, such as perchlorates,[26] these can be scrubbed or biologically remediated from Martian soils. Once improved by microorganisms, which add carbon and nitrogen, the Martian volcanic substrate is perfectly capable of sustaining crops:[27] on Earth, volcanic soils are some of the most fertile on the planet.

We can see, then, that obtaining oxygen, water, and food on Mars to sustain a human presence is possible. There is no showstopper to human life on Mars once the appropriate technology is mobilised. From that point of view, I think that a self-sustaining human presence on that planet is achievable, and it offers the greatest 'ease', if you like, of all the planetary bodies in the Solar System for human existence.

When we travel to the Moon or the asteroids, we have a more substantial challenge on our hands with respect to acquiring these basic needs.[28] There are no ice caps and glaciers on the Moon; however, there may be some sources of water. In the permanently shadowed craters of the lunar poles, there may be ice available, which can be extracted from the soil. Even across wider expanses of the lunar surface, water exists in the regolith as structural water in minerals that could be released with effort. The crucial ingredient here is plentiful energy, perhaps in the form of nuclear power, to bake or otherwise extract the water from the soil. That water would be electrolysed to make oxygen, or oxygen gas might itself be extracted from rocks by essentially releasing the oxygen atoms from the oxidised minerals in which it is bound. Here, too, energy is key. Like Mars, the Moon has plenty of basalt which should support plant and microbial life.

Similar approaches might be applied to asteroids. Some of them are much richer in volatiles such as water, in contrast to the desert-like features to be found on the Moon. Asteroids are made of rocky materials that vary depending on the type of asteroid one encounters, but most of them, apart from metallic asteroids, are essentially silicate bodies with some carbonaceous varieties. These materials have been shown to support microbial life[29] and probably would supply plants with their necessary nutrients, provided the plants are enclosed in a pressurised growth chamber of sorts.

Beyond our inner Solar System, we might consider locations further afield such as the icy moons of the giant gas planets. Jupiter's moon Callisto has been considered to be a favourable staging post for humans.[30] Its distance from Jupiter, which weakens the radiation hazard, and its ancient geology, which offers a stable surface on which to build a permanent settlement, make it an attractive place to consider a station in the outer Solar System. As an icy moon, the water ice material from its surface can be used to provide water and oxygen by electrolysis. Hydroponics and other technologies might be used to grow food, perhaps supplemented by nutrients and silicate material extracted from the icy crust.

We could go on a continuing diversion into the technicalities of how to subsist on the surface of other planetary bodies and the various contrivances that people have imagined for extracting oxygen and water and making food beyond Earth. Although these technical matters are remarkably interesting and I find myself easily led into contemplating them, more detail will not add to, or change, the basic observations that are relevant to the discussion here. The technically demanding processes that lie between raw materials and human existence in any location beyond Earth (and this applies to the free plateau of space as well, since if we want to live there, we will need supplies from nearby planetary bodies) create a high degree of interdependence between people and a much narrower division between life and death than in many environments on Earth. In particular, the continuous provision of oxygen places human beings in a relatively precarious position compared to those that breathe the essentially—from an individual human's point of view—limitless and freely available oxygen on Earth.

To give the whole problem a slightly more physical basis, we might say that there is a substantial kinetic barrier to liberty. Running through the observations that I have made above is the common theme that to live anywhere in space, there are kinetic barriers between the materials one needs to live and their provision—barriers that require machines, energy, and technology at levels rarely seen on Earth. In the case of oxygen gas, the oxygen atoms must be extracted from rocks, water, or, on Mars, from the atmospheric carbon dioxide by physical, chemical, or biological methods. Whichever method we choose requires the input of energy to overcome the chemical obstacles to the acquisition of the oxygen atoms. This applies to most other useful commodities.

It might seem that some of these technical requirements can be mitigated once resources have been acquired, through a simple recycling process; this observation is correct. The difficulty in obtaining essential commodities, and the apparently paradoxical local rarity of some of them in the vastness of space, such as water on the Moon, will drive a strong motivation to recycle as much as possible. From water to precious metals, technologies will be mobilised to ensure that as little waste as possible arises from extraction. However, although these recycling systems will likely be automated in much of their operation, they nevertheless still circumscribe a restricted life in which the efficiency and human interdependency of a closed circular economy of materials in a spatially confined society surrounded by a lethal environment is the norm. This observation applies in particular to the speculative long-term situation on an interstellar world ship where, even if the meagre raw materials that might be collected in interstellar space could be harvested, the lives of the occupants would be defined by recycling.[31]

The link between extraterrestrial physical conditions and tyranny

These conditions on other planetary bodies and in the open plateau of interplanetary space that I have described seem to open the door to substantial control over people operating in such unforgiving circumstances: in essence, to tyranny.[32] I state this as an assumption here, but one I will elaborate on and defend in the next chapter. I should emphasise that I am not necessarily attempting to implicitly convince the reader that conditions in space will inevitably lead to tyranny.[33] Indeed, the very potential for despotism may encourage the development of powerful social norms and devices to prevent it.[34] At this stage, my purpose is to point out that life beyond Earth is extraordinarily difficult and requires coordinated action in the production of water, food, and oxygen, substances which have never been collectively subject to simultaneous control by authorities in any known location on Earth.

Perhaps some definition of what I mean in using the word tyranny is in order here. Tyranny is in some sense merely the obverse of liberty, which I have already discussed: it is behaviour that undermines those things that we have identified with the pursuit of freedom. By tyranny, I include conditions in which individuals are treated arbitrarily and unpredictably, possibly violently, by the power structures they sit within, or conditions in which they live under domination and interference, causing the expression of individual liberties to be largely extinguished; I mean the forms of governance associated with this conduct. However, it should be emphasised that coercion can occur under an entirely predictable legal system set up to establish unyielding control over a population. In this case, arbitrariness may have all but disappeared, but the point is that the individual or a group of people has little capacity for any self-expression without fearing that they will transgress the oppressive laws established around them, and this can be a form of tyranny. Oppression can occur under a fully developed legal system as much as under lawless rule. I take these conditions as my operational concept of tyranny.[35]

There are other features of the space environment that add the same flavour to the pot. The lack of a breathable atmosphere does not just deny us oxygen, but indirectly propagates a series of other problems that interface with the conditions for freedom. As the atmosphere is unbreathable, or non-existent, there is no extraterrestrial environment in which people can walk or move unhindered. Now on Earth, we are usually forced to wear clothes and coats (not just because social mores require this but because otherwise we may freeze to death), but the requirement in space for a spacesuit to protect us against instant asphyxiation, radiation damage, and exposure of our bodies to low pressure places constraints on freedom of movement and brings forth necessary technical support to achieve this movement that is unlike anything on Earth. Even when we are ensconced in a pressurised habitat, we are escaping the accumulated onslaught of radiation damage. One could liken this situation to a snowstorm on Earth that pins one inside a house for several days, but in space the attack is unrelenting.

It is often difficult to comprehend the vast spatial scales that exist beyond Earth, emptiness so vast that one must wait about half an hour for light-speed messages of goodwill to travel between Earth and Mars, messages that travel between computers on either side of Earth in what, to the human mind, is an instant. These expanses of separation beget isolation of settlements with all its potential political and economic consequences; such isolated groups and markets have hitherto been mere abstractions in the minds of political scientists and economists. The extremities of isolation would be taken to their limits if we were to consider a world ship leaving the confines of the Solar System itself and heading to a distant star. Thus, isolation is mixed with extreme limits in freedom of movement and added as another layer of restriction to those that we have identified, with all of the social consequences.

Successful tyrannies in space may stultify liberty on Earth; successfully free societies in space can vivify liberty on Earth

These, then, are the conditions under which human society will emerge in the extraterrestrial environment, and although the difficulties and challenges they present will surely be reduced by our technological progress, they will not disappear; space is inherently lethal to humans to an extent that is not the case in most places in which societies have emerged on Earth. That is an elementary fact. But it is not a matter that should dispirit us. Nor is this observation necessarily pessimistic or dulling of our ambitions to build a space-faring society. Rather, in recognising and accepting this reality, we may be better able to fabricate institutions that direct the reality to its best ends. If the conditions of polities beyond Earth turn out to be hellish, then we have done our best to assuage them, and we may still achieve happiness and success in those societies. If the naturally altruistic capacity of humans when collectively confronted by extremes releases camaraderie and a blossoming of extraterrestrial liberties previously unanticipated, then so much the better, and we can allow the barriers we put in place to prevent tyranny to lie dormant and unused. Even so, we could still wonder why all this would concern us. Isn't despotism something for future space travellers? Why should we care on Earth?

It may be the case that the construction of different forms of society in space will influence the conditions for certain polities on Earth. If the Solar System is filled with illiberality, a scattering of settlements comprised of collections of slaves under the tutelage of despots, is it not the case that compared to them, any form of government in the oxygen-filled atmosphere of Earth will look benign? And is it not possible that this comparison might be actively used to shore up and give justification to despots on Earth? Is it also not a worry that liberal regimes and their populations that stare up into the night sky and see the void filled with tyrannies will feel insecure, sitting at the bottom of a gravity well under the all-seeing eyes of extraterrestrial autocracies? Even the smallest scattering of settlements beyond Earth may have this effect; it does not require parity between the population of off-world settlements and those on Earth to generate this shadow. Those who command the high frontier, who ply the upper reaches of the gravity wells, can exert a disproportionate influence on Earth.[36] The construction of the conditions for freedom beyond Earth is not ephemeral, or a fascination for a few academics who like to ponder the future; it may determine the very future of freedom on Earth.

Turned on its head, might we also observe that an energised effort to produce liberal societies in extraterrestrial conditions so extreme can lead to new technologies and mechanisms to bolster freedom on Earth? Might fresh ideas, approaches, and insights into the human condition that could improve liberty be formed? Could it be possible that if a group of people can generate a free society in a place where utter interdependency is required to even breathe a lungful of oxygen, this might bring shame on those who apply despotism to their population under the breezy oxygen-filled atmosphere of Earth? A Solar System of societies that have, through considerable effort and human ingenuity, produced the conditions for freedom might act as a measure against which terrestrial societies should aspire to improved conditions, removing the excuses for tyranny and pressing them to measure up to the achievements of their extraterrestrial brethren.

Liberty beyond Earth is not just a concept for another place. The sheer scale and extent of the space frontier compared to our tiny world, a small silicate fragment a mere 13,000 kilometres in diameter, defines an unparalleled and expansive vista for freedom. The direction that liberty takes beyond Earth may not only define the conditions under which the denizens of the space settlements must live, but it may also set the tone for liberty for humanity for centuries and millennia to come.

To give ourselves the best chance of thwarting autocracy on the space frontier, we must first use the realities that I have laid out in this chapter to explore some of the sources of, and encouragements to, tyranny and from where these threats might come. This is my next task.

2

THE CAUSES OF EXTRATERRESTRIAL TYRANNY

Space offers liberty, but tyranny from within the political body

Escaping from many thousands of years of settled human society, with its entrenched ideas, established viewpoints and traditions, and restricted habits, battered and given form by long periods of conflict and peace, is one reason for harbouring a desire to leave Earth. The promise of a new frontier, a place of freedom and of personal liberation, was what drove many to embark on ships destined for the New World in the seventeenth century. So, too, outer space offers that hope for some. That, no doubt, explains why many signed up for a one-way trip to Mars[1] when they were presented with that (ultimately abortive) offer.

I have no doubt that there is some truth in this view. It seems to be a matter of historical fact that there must be something to motivate that outlook. Unlike the Americas, we know that there are no indigenous intelligent civilisations elsewhere in our Solar System, no human moral quandary to grapple with. The Moon, Mars, asteroids, and the vast spaces between them lie open, without pre-formed political or economic order; in that reality lies some objective liberation. These new experiences in space will generate novel cultures, experiments in social organisation, and new technological know-how that will benefit the occupants of these settlements as well as their founder populations.[2] I also accept that there are many ways in which the existence of humans in the space frontier might bring forwards strong new forms of human sympathy and cooperation in such unknown conditions.

I have little to say on the matter of liberty that derives from escaping Earth's social conditions, other than that 'new experiments in living', as John Stuart Mill would have had it, will surely spring forth from it.[3] However, we should also realise that this efflorescence is not a foregone conclusion either, or that if it does occur, it will not come with unwanted or unintended side effects. Others point out the possibility of a darker outcome to the human mastery of space,[4] including some loss of our humanity in occupying such alien limits.[5]

There are undeniably other ways in which the settlement of space will manifest important forms of liberty. The spread of our civilisation to multiple planetary bodies may afford us some insurance policy against certain types of extinction, particularly if those new societies are self-sufficient.[6] We could gain some robustness against extinction caused by an asteroid or comet impact. An object a mere ten kilometres in diameter, similar to that which is thought to have caused the extinction of non-avian dinosaurs and many other organisms at the end of the Cretaceous geological period, could destroy Earth-based civilisation or render it close to extinction.[7] Yet, spread across the Solar System, human society would be largely immune to total annihilation in this way; it would have achieved a form of collective liberty: the freedom from extinction.

The vast resources of space could alleviate resource scarcity on Earth by providing us with new materials and energy.[8] Access to the metallic and platinum-group elemental resources of the asteroids and the possibility of gathering solar energy and beaming it back to Earth without the frustrations of cloud cover are just some of the possibilities that we might realise. And then of course the expansion of the human population itself into space could release us from the restrictions of the two-dimensional surface of Earth and allow for human numerical expansion into the Solar System, maybe eventually to the stars, with further benefits to the scope and potential of our civilisation.

And let us not forget the sure freedoms that will emerge in other areas of human activity. Confronted by the banality of climbing Mount Everest or crossing Antarctica, feats which must now be accomplished anew with obscure permutation of means, the space frontier offers us genuinely new ambitions in exploration. The first circumnavigations of the Moon or Mars; the first ascent of Olympus Mons on foot; the first crossings of the Martian poles unsupported; the first mountaineering treks across the methane peaks of Pluto. By expanding our horizons of human exploration to the edge of the Solar System, a new vista of 'firsts', which may be accomplished alongside scientific studies but will be as much about the sheer thrill of human stories and limits, will open up. This is a form of freedom, a released liberty in the expression of human accomplishment and daring-do.

And then there are the vast potentialities in art and culture in all its myriad forms, from painting to theatre, that will be given new impetus and raw extraterrestrial material on which the human mind can act to produce a florid output of artistic expression.

All of these strident and spectacular possibilities for liberty and human free expression, and the blossoming of ideas, that could emerge in space excite me enormously. However, these are not the focus of this work. The reason is not that I think that they are insignificant. The potentialities that I have raised might well turn into a motor of quite some considerable impetus for innovation and change that may redound to Earth and vivify our own ideas and systems of government, science, and culture and our technological capacity; I do not underestimate the potential—space offers new freedoms unbounded. I will certainly have something more to say about some of these matters later.

This book should not be taken as a denial of these forms of liberty; some of these, such as freedom from extinction or resource limitation, might be said to be liberties of a previously unseen and superlative scale. These are forms of liberation that are not merely for the individual or confined to a local group but encompass our entire species and perhaps other species from across the tree of life that journey into space with us. So let me emphasise the possibilities for forms of liberty beyond Earth that have hitherto been mere science fiction but might emerge into reality in the decades and centuries to come.

My reason for not giving these matters focus in this work is because I am more interested in identifying, and suggesting ways to combat, a problem that I think is a greater challenge to the human spirit and future: the emergence of tyranny on the space frontier. If escape from Earth should bring forward a remarkable flourishing of human creative capacity and freedoms, then I think we should all be happy about that, and I am sure we would all welcome it; any benefits that emerge from human progression into space will be a net gain for our civilisation. Alternatively, if the specific possibilities I have just raised do not come to fruition, then our position will be no worse off than we were before, apart from, of course, the resources we placed into the venture. We will just have to see whether these benefits accrue and in what form they emerge.

Yet, if despotism should emerge in space, then the consequences for those who move into this frontier, with their great hopes, could be profound, and that negative outcome may stymie, foreclose even, humanity's adventure to the stars, perhaps leading us to a restricted future on this small, irregular ellipsoid of rock we call Earth. Not least, as I have observed before, these autocratic societies may have a cooling effect on Earth-bound liberty as they look into the depths of the gravity well below them with illiberal ambitions. So the success of tyranny on the space frontier could have far-reaching consequences for our aspirations as a species and our practical achievements. A multi-planet species of cosmic tyrannies is not enticing as a long-term vision for our civilisation. Why should we think that such an outcome is possible?

The extraterrestrial environment is the handmaiden to tyranny

Although an escape from Earth may beget liberty, the greatest concern for the space settler, I believe, is the emergence of tyranny from within the very society that has achieved this liberated state beyond its mother planet.

In the previous chapter, I outlined some of the ways, both in degree and categorically, that all extraterrestrial environments impose physical conditions on collections of humans that are more extreme, in their collective effects, than any environment on Earth. Ranged against a society in space, surrounding it on all sides, are conditions that the society must confront.[9] There are fundamentally two ways it can do this. First, it can regard the environment as something to be accepted and the restrictions in movement and the efforts needed to survive within it as something to be submitted to. We might observe that humans who live in the extremes of the Canadian High Arctic, for instance, have this type of approach to their extremes: the environment is a reality to be dealt with and lived in concert with, not to be fought as something alien as such. A second approach is to treat the environment as some sort of enemy and to take on the collective attitude of a group engaged in a war against an oppressor that seeks to destroy the settlement. Society becomes an entity involved in a constant struggle against lethality, and the resources of the settlement are mobilised and aligned against it in a manner more reminiscent of a society at war, with all of its social implications.[10]

Whichever attitude a society adopts, the conditions of the environment do not change, and the required effort to survive within them imposes restrictions on movement and the mobilisation of resources that place heavy demands on those who live in such extremities. Now, this all-encompassing danger will no doubt inspire camaraderie.[11] I have worked in Antarctic stations and Arctic field camps and know the sense of common purpose and mission that emerges in a group of people subjected to physical extremes. However, I would note that these encampments are usually temporary for their occupants, rules are restrictive and taken seriously, and one distinctive aspect of acquiescence to these rules is the knowledge that one has not been assigned to spend many years, or the rest of one's life, within them. Nevertheless, it remains the case that I do not repudiate the claim that such extremities can breed kinship and that a sense of liberation can emerge from expressing one's skills and capacity in such a collective; a feeling of belonging can emerge from integration into a powerful community. In the case of a more permanent collection of settlements, this is well documented for the early years of the kibbutzim, a situation that was characterised by challenging physical conditions and isolation, not unlike extraterrestrial environments.[12]

The forces that concern me here are ones more detrimental to the health of society and its component individuals. Living at the limits of existence makes several phenomena possible, and I wish to discuss each of these in turn.

The first problem we might note is that environmental extremes lead to physical restrictions that are not so much an abridgement of liberty as an antecedent of social restriction. One's incapacity to fly without a wingsuit on Earth is not a denial of liberty and neither is the impossibility of walking out onto the lunar surface without a spacesuit. Locke, in considering the use of law to direct people to their best interests, had said: 'and that ill deserves the name of confinement which hedges us in only from bogs and precipices'.[13] However, being surrounded by an all-pervasive metaphorical bog, the need to wear a spacesuit anywhere on the lunar surface will restrict a basic ability to move freely in many everyday tasks that can bring on a sense of social claustrophobia not often experienced on Earth. Even in the Antarctic and the High Arctic, although the winter months can severely check movement, there are many days in these locations when one can wander outside with a large degree of freedom, breathing the oxygen-filled atmosphere without the fear of instant death. Although the necessity for action to escape lethal conditions is not technically a curtailment of liberty, the social consequences that arise from it can lead to this condition, or, perhaps more pointedly, they provide an excuse for others who seek power to use these dangers to promote such a social culture, or even to fabricate laws that pursue this end.

The need to impose restrictions on the use of spacesuits and the regulations required to ensure the safety and operation of critical oxygen, water, and food-producing machines in these suits and the pressurised habitats in which people live will yield a network of rules and regulations that have the side effect of creating a generally more delimited social condition. Although these regulations might be necessary, reasonable even, it may be difficult for individuals to divine this need, and hence a society may emerge in which a sense of coercion is more strongly felt. I have illustrated this notion in Figure 2.1, which shows the concept that on Earth the borders of laws and rules that prescribe human freedoms may be much wider on account of the physical environment that accommodates them, but within the extraterrestrial environment, a less liberal social organisation may be hemmed in by necessary restrictions that nevertheless lead to fewer liberties.

Now if these more curtailed liberties are necessary because of the environment, then we can legitimately claim that they are not restrictions as such, but realities of existence; but the situation they lead to is a system of all-pervasive rules that must be upheld. And to achieve this, an extensive system and culture of bureaucratic enforcement will be fabricated, which becomes part of illiberal governance.[14] In the extraterrestrial situation, people may feel they are in a less liberal state, although, as illustrated in Figure 2.1, in actuality the extraterrestrial authority may be more liberally inclined than a terrestrial state because it has done everything it can to stretch the rules and laws that circumscribe society to the maximum extent possible. To put this another way, an extreme libertarian society beyond Earth, pursuing the most laissez-faire vision of society it can muster, could still end up

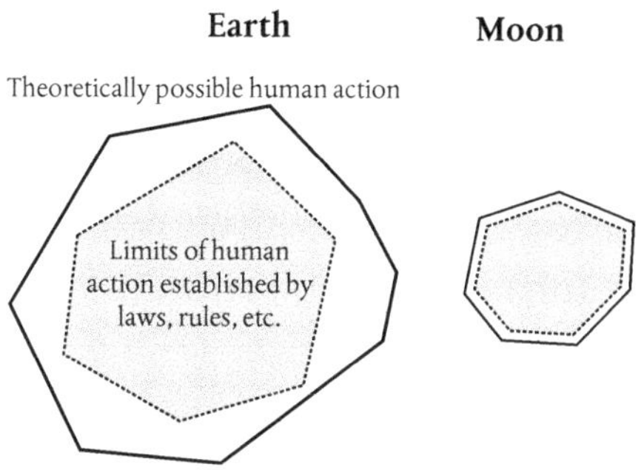

Figure 2.1. Limits of human action beyond Earth. A diagram illustrating the possibility that in a temperate environment on Earth, where the availability of oxygen, water, and food allows for a wide range of human action circumscribed by laws, rules, etc., a governing authority can potentially be less liberal than in an extraterrestrial environment (e.g. the Moon), where the governing structures might do everything they can to expand the laws and rules to allow for human action to the maximum extent possible. Nevertheless, the extraterrestrial society ends up more restrictive to human action than on Earth because natural extreme conditions limit this action (adapted from Cockell, 2015a).

more coercive than some societies on Earth that are considered tyrannous by libertarians.

The second phenomenon that living in extremes encourages is that in any environment where the possible actions of individuals are reduced, there is the likelihood that these limits are not only represented in more restrictive rules and laws, but also in the custom of restrictions. Perhaps a personal anecdote is useful here. When working in Antarctica in the summer of 2000, each of us was assigned by rota to a day when we would help the kitchen staff. As part of my scientific work there, I planned to carry out an experiment to study the damage caused by ultraviolet (UV) radiation to microorganisms over a 24-hour period. This involved taking measurements every 30 minutes over the 24-hour period, and it required a clear sky so that there was maximum UV radiation and a clear diurnal cycle to be able to discern the effects of the radiation over time. However, on the day of my assigned kitchen duty, the sky was utterly clear of clouds, so I informally agreed to swap places with the person assigned the following day. Unfortunately, I did not write this on the rota, and the appearance of a different person that evening caused consternation to the kitchen staff. Another member of staff, unconnected with the situation, felt the need to weigh in on the importance of sticking to the rota. Obsequious apologies and deference were needed to cover the cracks of this transgression, even though the task had been satisfactorily completed.

This trivial example, and I have witnessed many similar examples in polar environments, illustrates the point I wish to make. In this case, a system to ensure kitchen help had been set up and its subversion was seen as a threat to the spirit and unspoken rules of the station, whose existence in the extremes of Antarctica depends on conformity. Helping with kitchen tasks is not specific to Antarctica: this might have occurred anywhere, but it is emblematic of many such rules and boundaries that emerge in places at the limits of human existence. The extremes of Antarctica give these rules a force and authenticity that in any other location might seem, and be treated as, petty.

And this leads me to a related point that is paradoxical but nevertheless at the core of how power manifests in extreme conditions. Faced with an existential threat, an authority has the opportunity to do one of two things for any given situation in which an individual or group of individuals makes a claim on its power.

First, it can treat a claim to some type of freedom as irrelevant or trivial in the broader context of the lethal threat that everyone faces; minor 'rights' become subordinate to the greater common need to succeed against the monster that may engulf the settlement.[15] In this way, individuals can be presented as disingenuous or irresponsible for making apparently 'trivial' claims when the governing power is trying to grapple with challenges on which the whole settlement hangs. Livy had observed a similar phenomenon in Rome. Although in the Roman context this was brought on by the need to fight enemies, not the extremity of the physical environment, the general culture animated in the populace was exactly the same: people were drafted into the army to fight foes but also 'to give them no chance to draw breath in the City nor leisure to take thought for liberty or to gather in assembly'.[16] External threats, whether invading armies, Antarctic dangers, or extreme extraterrestrial radiation, are a ripe excuse to abridge freedoms. Under threatening external conditions, the authority can more arbitrarily choose which rights are to be accepted and which are to be quashed, even demonised, according to its own agenda in the settlement. When the potential death of many people is a counterpoint to any argument for freedoms, it is a relatively easy matter to make those claims appear frivolous.[17] A common enemy is an easy way to demand the suppression of otherwise dissenting voices.[18]

Second, the alternative and directly opposing strategy, as my Antarctic analogy illustrates, is to transform a trivial transgression into a fanfare about the higher demands and necessities of a settlement; even the most superficial transgression can be raised to public obloquy. The effect of this is to cower the population who now fear that any small indiscretion will place them in conflict with the settlement as a whole. These are the distortions that are possible, made easier, by the extraterrestrial conditions; at the risk of belabouring the point, they do not guarantee a despotic regime will emerge, but they increase the range of levers on offer to extraterrestrial authorities to suppress freedoms.

In potentially deadly environments, those who manage institutions do not want to be seen to be running a lax regime that could threaten lives, and those within the system do not wish to be caught out as the incompetent who threatened everyone else. The condition depicted in Figure 2.1 leads to two essential outcomes. First, a network of officialdom charged with ensuring safety emerges alongside an entourage of willing volunteers who wish to be seen to be upholding the system. They use the networks of officialdom to demonstrate their loyalty to the bureaucracy for numerous personal reasons, which might include promotion prospects, the love of power games, the desire for authority, a desire to show high moral awareness, boredom, and so forth. In other words, the required rules generate an all-encompassing web of coercion which ultimately plays into the hands of those running the settlement.

This network of officials exists whatever type of governance structure is chosen for the settlement. I have previously called this an 'oligarchic core',[19] but it is not necessary to agree with this definition to recognise that however a society is run, there is always a cohort of officials and their allied functionaries who are not subject to popular censure or democratic mandate and whose job it is to enforce rules.[20] This is not merely a police force, but includes health and safety officials, construction standards managers, etc. They are not necessarily despotic, but they constitute a spider's web of personnel that act as the conduit through which power permeates a society. Beyond Earth, their potential is legion. We might imagine oxygen, water, and food production and safety officials; those charged with the safety of spacesuits, pressurised rovers, and habitats; those responsible for safety of people in the field and travelling beyond the confines of the settlement; officers to quantify and watch dust incursions into the settlements; radiation monitoring agents, etc. The sheer variation and extent of threats to survival makes possible, perhaps even inescapable, the expansion of this network of potentially unelected yet immensely powerful people on whom existence depends. Here again, our task must surely be to mollify the potential excesses of such a network and to put in place mechanisms to prevent it becoming an instrument of all-pervasive control. To this task I will turn in the next chapter.

The second essential problem that results from extensive networks of regulations that the extraterrestrial environment demands is the slow consolidation of a culture of conformity through the settlement.[21] No one else will swap places on something as trivial as kitchen duties, for fear of similar public repercussions. Thus emerges a dulled sense of individuality, an unwillingness to not only challenge rules, but even to flexibly interpret them in a non-destructive way. In short, extreme conditions breed conformity, an extraterrestrial tyranny of the majority or custom.[22] Again, lest the reader misinterpret this to be a singularly negative outlook, I would point out that conformity also breeds safety: no one wants a Mars station full of out-of-control renegades. The matter I wish to draw attention to, however, is that the tyranny of the majority emerges from within the belly of a collective of people submersed in extremity. The liberal expression of countervailing behaviours and attitudes is stymied and this rigidity seeps through the entire system, ultimately giving succour to the power ambitions of those who run the settlement, particularly if those individuals have a tendency towards that behaviour.

Ultimately, even the tyranny of the majority may seep away. Minorities themselves, losing any interest in challenging the authorities, retreat into conformity, and what remains is a society of flat homogeneity of opinion, where even a single expression of dissension on any matter is an anomalous phenomenon that immediately attracts the collective wrath of the population. Faced with the prospect of espousing views that are publicly treated as courting danger for the whole settlement, it becomes easier to conform than to think. Self-expression through deliberate democratic means cannot necessarily save a society from this fate, since the demand for equality, manifested by the democratic principle itself, may give force to these expressions of conformity and majority power, driving society towards an insipid sameness. To escape this fate, the deliberate encouragement of a variety of voluntary associations that dispel conformity must be attempted alongside democratic processes. I elaborate these means later in this discourse.

Conformity is destructive in that it hides possibility. In a society faced constantly with the exigencies of survival and success in a lethal environment, the social conditions should be encouraged towards a state where heterodoxy can be used to spawn innovation. To put this another way, out of the need to survive emerges a compliant culture and set of behaviours that are antithetical to the very intellectual qualities needed to enhance the chances of long-term survival. Conformity can be self-destructive.

For evidence of this central feature that I claim will shape most of the conditions for tyranny, we might examine small settlements in some locations on Earth. Although not nearly as extreme as the extraterrestrial case, the Scottish islands of the Hebrides are instructive in their lessons for the sociology of long-term settled communities in isolation and environmental extremes. Some of them have hosted populations of around 100 individuals for many hundreds of years, akin to a small extraterrestrial self-sustaining settlement. I have spent eight years studying these communities with the purpose of understanding their lessons for the extraterrestrial case, both through the literature of their history and through conversations with the islanders in the present day. Here I want to point out some evidence in support of the matters I have raised.

In regard to the small community of about 100 people on the island of Colonsay, one of its inhabitants observed the following:[23] 'The familiar goldfish bowl, which seems so friendly and comforting most of the time, can suddenly feel suffocatingly small . . . Creeping peer pressure could come out of a clear blue sky.' On the island of St Kilda, another Hebridean island now no longer occupied by an indigenous long-term community:[24] 'Many people . . . would regard life in St Kilda as anything but ideal . . . because of the absence of opportunity for individual originality and creativeness, which, though to some extent dependent on the firm support of a culture, are easily smothered by a cultural pattern which is too strong . . . Undeniably there was little room for individuality.' Put differently: 'The common survival was the prime concern and individual betterment was an idea scarcely conceived of.'[25] The drive for conformity on these islands is accounted for partly by the need to cooperate in extreme environments (particularly in the winter) and partly by the fact that small populations being localised to a small area and in constant proximity find that deviation from norms tends to elicit suspicions of people getting 'above themselves'.

Aside from Scottish islands, powerful incentives for conformity are reported in small communities in very different settings, such as the kibbutzim. This latter context is of special interest because these communities in the deserts of the Middle East overtly prided themselves on a socialist vision of humanity, an acceptance, in other words, of a deliberately more collective social organisation. Yet even here, the drive for conformity has been reported to be intense. In an examination of the social conditions that emerged, Spiro noted: 'Hence, the sheer frequency of interaction, brought into being by the nature of communal living, as well as by the intensive kibbutz schedule, keeps one constantly exposed to the public eye . . . Should a chaver [a person in the community] violate the group norms, it would not only be known in a short time, but he would be openly criticized for his behaviour. Such group censure, informal though it is, is highly effective in a small community, and would render his existence in the kibbutz untenable.'[26]

The collective result of this tendency to conformity is something akin to Rousseau's 'general will', a type of agreed social order given force by formal rules and custom into which all individuals submit.[27] As these boundaries are necessary for survival and their dismantling could increase the chances of significant damage or death, these rules, whatever their extent, can be presented as reasonable; they provide for the liberation of the individual from the lethal extremities that constantly knock at the door of an extraterrestrial habitat. Even if an individual is not aware that they have erred, so long as their error is detected by someone else, then they can be 'forced to be free'.[28] In correcting their actions, they have been saved from themselves. Instead of a coercive act, the imposition of control over an individual becomes a measure to advance their freedom, a benevolent act of liberation. Although this state of affairs that I have described could emerge anywhere, and indeed one might even claim that elements of it exist in any society, the point is that the intensity of the extremes in extraterrestrial environments can make this culture total and unrelenting in its influence on the social order.

The style of this changing character of conformity will inevitably alter if a settlement grows. This changing culture will influence the way in which we seek to instantiate liberty, to the greatest degree possible, within a settlement. Therefore, the nature of this changing structure of liberty is important to identify.

To begin with, small groups of people give more opportunity for a person's voice to be heard. As a settlement grows, anonymity is more possible for the individual. Put simply, there is a wider range of people one does not know well, and this weakens the peer pressure to conformity experienced in a lesser population. Yet this larger structure makes it less easy for a voice to be heard and it makes it simpler for those governing the group to implement a range of measures that are less immediately obvious and open to democratic censure. Even if the political system is organised around a representative democratic system allowing for means of personal political expression, everyday political expediencies, decisions, and actions are less visible than they would be in a group of a few tens of people. In this case of a representative democratic system, a range of checks and balances within the system can be used to ensure that these politics are constrained in a way that is beneficial for the freedoms of the settlers. I will have more to say on these matters in the next chapter, but here, from the point of view of foes to liberty, I intend only to point out that the growth of a settlement inevitably causes the dynamics of those things that do represent a threat to different forms of liberty to change. There is no universal 'anti-tyranny' recipe that can simply be applied in the extraterrestrial environment at all times, but there may well be common general approaches that can be identified for any given general size of a settlement.

Whatever the size of the settlement, from the conformity-inducing extremes a severe power asymmetry can easily emerge between those running the settlement (and their technical supporters) and the rest of the settlement. Although we may wish for a more collective view of society in space extremes, where the individual expresses their capacity and contributions more forcefully by belonging to society, the extraterrestrial environment forces the body of people towards a state of affairs that lends itself to further decline into collective despotism. Mechanisms and safeguards must be incorporated into society to prevent this deterioration.

Worse still, this projection of power can become arbitrary. The oxygen machine of a troublemaker might be subject to emergency repairs which are actually technically unnecessary, to remind this person of their precarious existence: deadly environmental conditions become an instrument for capricious authority. In any matter from which death may come on swift wings, this unpredictability can be justified as the need to act immediately in the interests of the wider polity. Apologies for coercive interventions can cynically come later, but only after the point has been made. Thus, where extremes justify quick action, it is the potential arbitrariness of this power, not merely its extent, that should trouble extraterrestrial settlers.

In addition to strengthening and giving lifeblood to a potentially more overbearing authority, extremes will shrivel and diminish the sphere within which individuals can pursue their own ideas of the good life. Not only do a network of substantial regulations and rules make an individual more conformist, but they also reduce the scope of free action and reduce the confidence with which individuals feel the capacity to express their own views and activities, even when they are physically alone. In the lexicon of Isaiah Berlin, we could say that the extraterrestrial environment has the likely effect of shrinking individuals' 'negative liberty',[29] that private domain within which an individual is free of coercion by the state or its analogous extraterrestrial equivalent. Here, too, we might observe that this lesser field of action of the individual may simply be an ineliminable part of the necessary limitations on individual aspirations when people are corralled under lethal conditions. The salient point is that the extraterrestrial environment pushes with much more force in the direction of a diminution of this private sphere of activity, and with that pressure comes the excuse and the capacity to utterly destroy it.

The dystopian end state of the expansion of conformity and the restriction of negative liberty is a group of contended slaves doing the bidding of their extraterrestrial masters, mere functionaries with no internal spirit and expression of individuality, having given up this sense of self voluntarily to secure their own survival.[30] It is this conclusion that the extraterrestrial environment convincingly encourages us towards, and it is against this final state that we need to construct the social, political, and economic mechanisms to guard against.

Nor should we convince ourselves that the mere presence of some meagre liberties is proof that this unfortunate end state has not been reached. It is perverse, and hopefully inaccurate, to make comparisons between extraterrestrial settlements and concentration camps, but Frankl's account of his period of internment in the Auschwitz-Birkenau and Dachau camps, among others, show that even when one is reduced to a complete state of brutal dependency with everything one possesses having been removed, one can still experience forms of liberty. He observed:[31]

> Man *can* preserve a vestige of spiritual freedom, of independence of mind, even in such terrible conditions of psychic and physical stress. We who lived in concentration camps can remember the men who walked through the huts comforting others, giving away their last piece of bread. They may have been few in number, but they offer sufficient proof that everything can be taken from a man but one thing: the last of the human freedoms—to choose one's attitude in any given set of circumstances, to choose one's own way.

I think the pertinent observation—or perhaps warning—here is that an authority can claim the existence of freedoms, and the people under its regime may even feel these freedoms, under the worst possible human conditions; yet the mere existence of some forms of liberty does not, and should not, provide some basis for a claim that the society has goodness underpinning it. Frankl even draws attention to the 'independence of mind', one of the much-vaunted capacities encouraged in free societies, present in the huts of Auschwitz. In the extraterrestrial case, we are not trying to protect some minimal level of human freedom which can be identified as such, since this was possible in Nazi concentration camps: we are in pursuit of a comprehensive set of liberties that elevate the human condition to one of non-domination by others, particularly where that domination is arbitrary, and a capacious self-expression that is worthy of the human mind within extreme physical settings.

The challenge in the extraterrestrial environment is not merely to fashion a society in which some semblance of individualism and personal self-worth is possible, but to manage a continuous and inherent conflict, a paradox if you will.[32] On the one hand, the bland, monotonous, and dangerous characteristics of the external environment give more pronounced encouragement to the expression of individualism, an urge to express one's creative distinctiveness; on the other hand is the sheer demand for collective coordination, the characteristics of which we have already examined. These two facets of society are constantly in tension, permanently pulling each other apart. If the former should win, the society becomes a collection of anarchic uncoordinated individuals who will surely eventually fall foul of the conditions in which they exist. If the latter should win, the society collapses into something akin to a slave state. It is not that this tension will be new; let us recall Thomas Hobbes's observation of the difficulty of navigating between too much liberty and too much authority without getting wounded. Rather, it is that the need for individual self-expression and the need for collective control tug against each other with more appreciable effect. An over-exaggerated worship of the individual can result in the impulses of romanticism that decry the rational order and organisation so vital to the scientifically and technically driven needs for human survival in space.[33] There is no need to get drawn into a discussion about whether the Enlightenment offered us objective universal values of human organisation or whether those values are highly dependent on the individual, as the era of romanticism would emphasise, to recognise that both outlooks strive towards a sphere of individual activity free of coercion and arbitrary rule.[34] The extraterrestrial environment degrades this capacity for individual expression in whatever way we interpret it.

This view of people and societies should not be misinterpreted as a suggestion that there are not powerful forces *against* tyranny in space. Just as I have illustrated the ways in which settling space may liberate humanity in a number of activities, within extraterrestrial societies themselves there will be powerful counter-currents against autocracy. Given a choice to leave a tyranny, people will tend to do so, and extraterrestrial authorities, perhaps fearful that they could lose people critical to their survival, may be inclined to treat them well and keep them content. A small group of people gathered in a potentially lethal environment have very strong motivations to achieve comfort and social concord and, as I have already pointed out, to draw upon a strong sense of camaraderie, as we observe in terrestrial polar stations. There are other drives, such as the multigenerational ambition of space settlement, that could encourage deeply altruistic outlooks.[35] I do not deny powerful forces against autocracy; my purpose here is to assume the worst with the intention of identifying how we might achieve the best.

Extraterrestrial authorities will face a powerful struggle between spontaneous order and central planning, with a tendency to the latter

A consequence of the aforementioned conditions that I want to draw attention to is that they will stymie, or at least weaken, the confidence people have in the ability of spontaneous order to solve society's problems. In the next chapter, I will suggest how such voluntary ordering in many of the activities of a settlement might be encouraged, but here I want to emphasise a few aspects of this phenomenon.

The reliance on the spontaneous ordering of society to solve its multifaceted problems is a matter that defined a vigorous debate in the twentieth century, particularly between individuals such as Hayek and von Mises, who both favoured the role of spontaneous ordering as a means to address society's political and economic challenges (by spontaneous ordering I mean voluntary association, markets, and the role of individual choices in establishing the patterns of social development), and those who instead saw, and desired, a more expansive role for the state in planning.[36] We do not need to digress into a long discussion of this history, or in fact even express a specific opinion about it, to recognise that the lethal conditions in space will have a tendency to encourage a firm commitment and trust in planning.

Since no extraterrestrial authority will want to be seen to have allowed disorder in the settlement, there will be a tendency for them to meddle in a range of activities to stave off disaster, want, anarchy, or any other number of ills that they perceive to be related to a lack of planning. In intervening in this way, and entirely for beneficent purposes, the remedies may create unintended consequences that require further intervention elsewhere.

Quickly, the extraterrestrial settlement and its organisation become akin to a series of never-ending extraterrestrial fires that must be extinguished. In the minds of the authorities, the response to these problems must be more planning. Fearing for their lives, the inhabitants will also demand and support further planning and assistance. The idea that the solution to this entire situation is to sit back and allow all the problems to spontaneously order and correct themselves will be seen to be nothing short of criminal irresponsibility. And of course, any single disaster, particularly one that involves loss of life, which may occur on account of this decision to reduce planning, will be seen to be evidence of the anarchy and danger of spontaneous ordering and will be met by even stronger demands for central direction. In short, the population themselves will always show a favouritism towards strong direction and leadership and a disinclination towards a perceived fragmented and distracted liberty.

Thus we can see that, surrounded by incessant danger, not only is the extraterrestrial settlement likely to tend towards modes of central planning and away from allowing spontaneous organisation, but a ratcheting effect of yet more planning will emerge from the desire of the governing powers to be seen to run the community in a competent way, and from the way in which planning will beget the need for more planning as the whole edifice of social order becomes demanding and arrogated by the structures of this oversight.

This direction of travel would be reinforced by the fact that when the potential for death is much closer, it may seem that a larger number of spontaneous movements of society have a greater chance of colliding with the walls of lethality and extinguishing society. Can we rely on a group of people to see that the emerging order within their social collective is inexorably heading to disaster, particularly when by its very nature it is spontaneous, unplanned, and unanticipated? With such fears, there will always be a temptation to head off society from perceived fates that await it before it is too late, and with that motive comes the irresistible desire to impose upon people an all-enshrouding predictable plan.

It would be as if a blind-folded friend of yours has been placed in a corridor and told to walk to the end of it to retrieve some food and you are provided with the opportunity to help them out by giving directions. If they hit the walls on either side, then they are denied the food. If the corridor is very wide, then you might expect to set them off in the right direction and then sit back, fully expecting them to reach the food with minimal intervention. However, if the corridor is not much wider than a person, and you cared for your companion's hunger, you would be concerned at every deviation they took and you would be in a constant state of issuing commands to them to move left or right along the whole length of the journey. Such is the state of an extraterrestrial society as it moves forward through unlit and unknown social territory with disaster close by on either side, its governors and population issuing orders to attempt to keep it on the straight and narrow (Figure 2.2).

We might observe yet another phenomenon that can strengthen the power ratchet. Given such high stakes for any given direction of society, the stakes of course involving its very survival, there may emerge strong rivalries for power. These conflicts could arise because of the strength of feeling about desired directions and outcomes, which could involve human lives, and because of the temptations of the immense prestige in claiming to be protecting society from itself. The considerable consequences of disaster, and the accolades in emerging as the

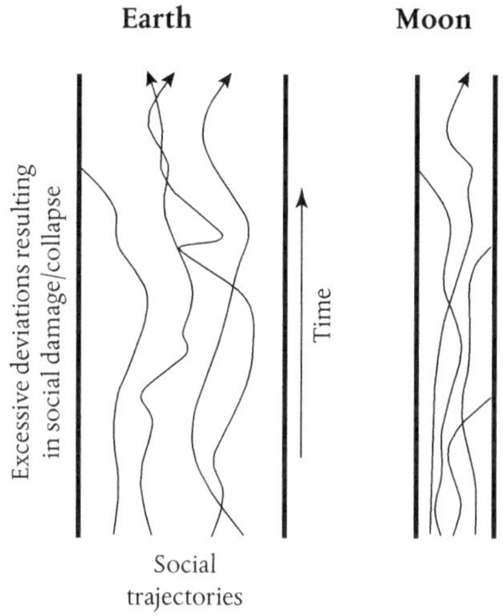

Figure 2.2. Spontaneous or random movements of a society that is constantly closer to potential danger can more easily result in disaster or catastrophe, encouraging the emergence of a forceful culture of political and economic central command, direction, and planning, which the inhabitants themselves may demand. In this figure, this is conceptually illustrated through a comparison between the Earth, where environmental conditions in many places on its surface may be more forgiving of social misdirection without causing cataclysm, and the Moon, on which deviations in social trajectory may more easily collide with destructive consequences.

saviour of society, have the potential to generate substantial jealousies between centres of power, in whatever form they are manifest, whether in individuals or organisations.[37] As centres of power seek to gain influence and power over rivals, there is an inevitable arms race within a confined environment, which can engender suspicions, subterfuge, and other undesirable and potentially highly toxic forms of social interactions. Power jealousies and rivalries are, of course, as old as society itself, but they are yet another element that can nudge a society towards unyielding and uncompromising forms of order and degrade spontaneous forms of human interaction that might produce opportunities for the coalescence and emergence of alternative centres of power.

This centralisation of power is not only unhealthy for political liberties, but it will seep inexorably into the economic sphere as planning reduces the number of economic innovations, a point I shall return to. The less spontaneous order is allowable, so too the less individual and group innovation, in all aspects of human thought. As extraterrestrial societies will begin in conditions previously unknown and uninhabited by humans, the unpredictability of solutions to the human condition, which Hayek famously drew attention to, will be even more marked in such environments than on Earth. Here we have a double-edged sword. The extraterrestrial environment encourages less spontaneous order, yet it also has an unpredictability that requires this spontaneous behaviour to spawn new innovations to solve a range of previously unencountered problems.

There is the simple truth that in extraterrestrial extremes more planning is likely to be needed than might be the case in locations on Earth, where there is no requirement to consider the manufacture of air, for instance. However, the relevant point here is that there are strong impulsions riven through the social order of extraterrestrial communities to move away from modes of human relationships that emphasise voluntary self-emerging order and towards patterns of central planning, even if such a community has consciously decided to create a society dedicated to forms of spontaneous ordering. An antidote to this is not to vilify planners, for their designs might be necessary, but, in realising this inherent tendency, to find means to enhance voluntary cooperation and spontaneous interactions to the greatest degree possible within the inescapable communal needs that society will have.

The traditional arguments for individual liberty are the basis of extraterrestrial coercion

The apparently necessary limitations on human action made real by the extraterrestrial environment are given potency by the very arguments traditionally used to defend individual liberty. John Stuart Mill's view is a case in point. His central doctrine, or statement, if you like, on the definition of the extent of individual liberty was as follows: 'That principle is, that the sole end for which mankind are warranted, individually or collectively, in interfering with the liberty of action of any of their member, is self-protection.'[38] On a planet like Earth, where the external environment is not instantaneously lethal, it is a relatively easy matter for me to enjoy movement and action in a wide variety of activities without being interfered with by others, or without my own actions impinging on others. Mill's need for self-protection, we are probably glad to realise, is rare. However, where death is just a cracked visor away, even on a thoroughly banal walk into the wilds of Mars for an hour, a variety of incursions on individual liberty can be justified on the basis that any errant conduct could result in danger or death and that the settlement, in curtailing the actions of people, is acting in self-protection.

Mill's formulation is a specific one to be sure, but the principle he adumbrates is not dissimilar to many general statements of individual liberty that circumscribe the notion that when others are threatened in some way, they have a right to violate a sphere of non-interference to protect themselves. Even before him, Hobbes was of the view that[39] 'all the duties of rulers are contained in this one sentence: the safety of the people is the supreme law'. The possibility that an action in a deadly environment might wound or kill leads to the arguments for individual liberty being used to restrict individual liberty itself. This contorted application of liberty need not even be cynical: it might derive from a genuine benevolence, but it will be a very real limitation on freedoms, and potentially an erosion of the very philosophical basis of liberty.

More devilishly, as all of these powers coalesce and envelop society, they may become so actualised in the culture that the settlement becomes enamoured of them.[40] Protected from instantaneous death, the population may even come to love their despotic rulers who give them their life, liberty, and protection against the external environment.[41] Alexis de Tocqueville noted this phenomenon in the attitude of the populace towards French kings before the Revolution, observing that this subservience was pragmatic because 'it was useful or could do you no harm'.[42] Yet if these extraterrestrial overlords secure your existence, there is every chance of a servility and dependency emerging from it that might well do harm to the existence of any form of liberty within a group of isolated individuals.

This problem of control and its defence through the time-honoured arguments for liberty that is manifested through the need for regulations within a settlement is also evident outside the physical infrastructure itself. Ranged underneath the authorities is a group of people who cannot easily escape from the settlement and whose safety depends on the rules and regulations that apply not just to the internal physical structure of the settlement but also to the capacity to leave it in spacesuits, pressurised rovers, and other vehicles. The extreme restrictions of movement that dictators on Earth must muster all sorts of arguments to justify to their people are delivered naturally by the extraterrestrial environment. A plethora of requirements, such as checking spacesuits, preventing people from leaving settlements because there are insufficient numbers of spacesuits and rovers, and monitoring people who leave the settlements, including their conversations, 'for their own safety' can be summoned to centralise power and create a structure of coercion around the blunt facts of the extraterrestrial condition. In this way, individuals and the settlement itself are being given the gift of self-protection, their liberty no less.

Beyond the settlement, and considering the situation of any group of people in space, we might ponder on another troubling ingredient in this noxious mixture: the vast spatial scales that lead to the relative isolation of settlements. Even on Earth, a determined despot often has little concern for liberal nations next door, but consider hypothetical political arrangements on Mars where nearby settlements may be absorbed with their own survival and help from Earth is months away. Even under the watchful eye of a government or corporation based on the same planet, let alone Earth, the managers of a settlement may have more free rein to indulge distortions of illiberal power with fewer likely repercussions if they are not integrated into a much wider polity.

Physical isolation and extremity immediately in the periphery of the settlement forestall the possibilities of leaving the settlement should an individual find its regime suffocating. Unless deliberate attempts are made to implement the right to leave, the default condition is confinement. Even if the means to leave is provided, this could be weeks or months away, leaving the individual at the mercy of the authority he or she has parted ways with. It is easy enough for an authority, confronted with an individual raising complaint about their situation or the predicament of the entire settlement, to offer to open the airlock door so that the individual may fulfil their desire to leave. This is, of course, a particularly misanthropic scenario, but it underscores the fact that from the point of view of a settlement authority, the two options on the table for a dissenter are to 'agree with us or exit the airlock'; they are coercively offered their liberty, which is death. The challenge for us is to attempt to conceive ways to minimise this culture.

To summarise the rather dour but I think essential preceding discussion, what we see here is a nexus of potentialities for tyranny converging in the extremes of space: the possibilities for coercive, and potentially arbitrary, power[43] that revolves around the provision of vital commodities such as oxygen, water, and food; the need for strict and sweeping health and safety regulations to ensure welfare within the settlement; the opportunity to transform the natural restrictions of freedom of movement into arguments for control; and the isolation of the settlement as a whole, which allows for power structures that are subject to less external oversight. Paradoxically, toxically, holding these plates together are the intellectual arguments for individual liberty itself. This fearful edifice does not guarantee the emergence of a despotism, but it facilitates its inception by creating an all-enveloping structure of power that can be pushed beyond the necessary limits to successfully survive. Even if we were to aim at a minimal state in the most libertarian vision of existence,[44] the extremities of outer space can make this politically minimal state overbearing. Indeed, it raises a fundamental political question on what the theoretical minimal state in an extraterrestrial environment is and whether such a state can ever achieve a size that comes anywhere close to that which a libertarian would conventionally consider minimal. It is from its perceived responsibility to protect a population from their demise that any extraterrestrial governing structure, of any political persuasion, can project a supreme authority that can be justified and asserted with omnipotent power.

Extraterrestrial isolation encourages economic tyranny

Running parallel to the political arena, but of course heavily embedded within it, is economic activity; I separate political and economic activity for the sake of convenience. Political territory I have considered to be the general conditions of human activity and management. The economic arena I consider specifically to be financial transactions and the sale and purchase of goods. Of course, the 'economic' system is intrinsically part of the overall organisation of a settlement, yet some factors are worth drawing attention to, as they pertain specifically to what people would regard as economic activity.

One of the most obvious characteristics of an extraterrestrial settlement is isolation from other markets and sources of production. As on Earth, economic isolation can potentially have a number of knock-on effects, including: a stymied market for goods which become mainly, or wholly, domestic; a restriction in choice for individuals who cannot easily leave and whose functions become critical if there are not alternative potential employees to take their jobs; and monopolistic control of certain industries when there is very little choice and those industries are essential for survival.

In essence, by separating a small group of people with their economic lives from other markets and labour, we potentially end up with an etiolated economic system, in some cases even a bone fide autarky; this restriction plays directly into the political opportunities for tyranny. As the settlement's choice of goods is restricted, particularly if this is in the matter of parts for vital machines, it may become necessary for the goods that are produced to be tightly regulated to ensure that no machines go short. The limited scope of production can force controlled rationing and, as with the political sphere, it may not always be the case that the individuals and groups charged with ensuring the safety of supply will merely expand their control to the minimum extent necessary to meet and secure those needs. Instead, economic restrictions encourage expansions of political power which are manifestly more justified when the consequence of failure is death.

Limitations in the means of production of certain goods may be joined by restrictions in the human capital. The incapacity to export or import large ranges of goods, a symptom of a lack of regular trade routes, will by default curtail human movement as well. Now the settlement may be faced with the potentially disastrous lack of key personnel in key manufacturing roles if they choose to leave or move employment. Out of this predicament emerges the need to control employment to ensure that machines are kept running and to prevent the deleterious movement of personnel within the settlement, which may disrupt a carefully managed equilibrium of production.[45] From highly economically restricted conditions, therefore, emerge the requirements for highly restricted political and social arrangements. The erosion of economic liberty by the extraterrestrial environment and its spatial scales that encourage isolation leads to the deprivation of political liberty. To achieve political and social liberties, we must also maximise the potentialities for economic freedoms.

Religion will either aid or mollify extraterrestrial tyranny

Often connected with the political and economic activities of humans are their various cultural façades. One of the most prominent of those is the diversity of personal and collective beliefs about higher purposes, manifested most frequently in organised religion. Faced with the hopeless deadliness of the external environment, it is easy to see how groups of individuals might turn to some form of spiritual escape. Against the conformity-generating extremes, this sense of belonging to a religious order might bring meaning to life; the potential to turn towards a 'higher power' that stands above the everyday banality of the lunar and Martian deserts, and the mortal limitations of humanity, may be an attractive possibility.[46] Furthermore, the authorities of an extraterrestrial settlement could be drawn towards the force and persuasiveness of an extra-human power that could be commandeered to justify their own power and authority over individuals. The mobilisation of the whole settlement against the environment, whose conditions are almost an embodiment of a satanic force, transforms submission to the proclamation of authorities into religious salvation. The merging of religious belief into the pragmatic goals of political power creates a theocratic potential the likes of which has never been witnessed on Earth. On a practical level, the manifestations of this reality would be the formation of a cult-like community where the individuals appear to willingly submit to this order, yet are effectively brainwashed into a submissiveness that secures their existence.

It is also undoubtedly the case that a powerful religious view may find a ready and enthusiastic audience in a group of people isolated from others and confined to a foreboding environment. A historical case in point is the Scottish island of St Kilda. On St Kilda, for as long as that community existed, islanders met every day; the Mòd or parliament was used to decide on plans for the day and resolve any community issues. Despite the participatory democracy, the island did fall prey to despotic influences, mainly through the vehicle of religious representation. After 1843, when the Free Church split with the Church of Scotland, it took over the parish of St Kilda. The man the Free Church sent to look after the island, the Reverend John Mackay, was a tyrant, and the description of the consequences is best left to Charles Maclean:[47]

> Only one of them managed to fill the vacuum created by Neil Mackenzie [the previous minister]—not with enlightened leadership, but with blinkered despotism . . . Shortly after his arrival . . . he established a vibrantly harsh rule . . . Services on Sunday at eleven, two and six o'clock were made to last for two to three hours each . . . Mackay preached long, repetitious sermons in Gaelic, which invariably included the same message of hell-fire and eternal damnation to all sinners . . . Children were brought to church at the age of two . . . conversation between the islanders was forbidden from Saturday evening until Monday morning . . . singing or whistling was a serious sin . . . children were forbidden to play games and even made to carry bibles under their arms wherever they went.

Mackay's rule occurred in a nineteenth-century context in which electronic communications did not exist. Today, his behaviour would rapidly become known on the mainland (although St Kilda is no longer inhabited by a permanent settlement). Nevertheless, what happened on St Kilda provides a generic lesson on the way in which an isolated population can fall under the sway of a determined individual, even when it ostensibly has a participatory democracy. The force of such individuals derives its influence from the fact that they appeal to a higher power to impose a dictatorial regime, making a challenge to the autocratic conditions a challenge to the will of a god.

In the extraterrestrial case, religious belief is not necessarily to be shunned. Regardless of what our views may be on the evidence for the existence of a god or gods, for some people, religious belief may provide the essential spiritual ingredient in their tolerance for the confining extremities of outer space; it is one of the many expressions of a freedom of conscience that is part of the collection of personal liberties to be protected and encouraged in space.[48] At the collective scale, a religious order provides another sphere of identity separate from the group running the settlement, contributing to the diversity of overlapping voluntary associations that do much to dull the concentrated power of single authorities. Against the vast and implacably austere conditions of space, the inner peace and sense of community engendered by belief may even give birth to new and intense forms of religion, which might range from a theistic belief to a type of pagan deference to, a worship of, the physical environment that plays such a thoroughgoing and imminent role in everyone's lives.

To alight on another terrestrial form of spirituality, a Buddhist self-control and detachment from the imminent danger of the outside environment might maturate; not only is an inner peace in accepting the impermanence of the human existence in the cosmic order—which is so present in lethality—a likely culture, but it might even help a settlement achieve a form of what can only be described as extraterrestrial nirvana. In other words, we do not have to hypothesise the existence of deities to appreciate that certain forms of spiritual structures, which we tend to collect under the banner of 'religion' on Earth, could provide social order beyond Earth, and represent a form of liberty from the perceived power of the external conditions for the individual. Moderated and applied in controlled doses, forms of spiritual commune could well play a positive role in securing the social cohesion of an isolated community.[49]

Yet, it is evidently the case that we could imagine the pull of religious beliefs becoming an instrument of tyranny that overwhelms a settlement, as it did on St Kilda. The modern-day facility of electronic communications, not available to the St Kildans, will not necessarily obviate such a despotism. The enormous spatial scales of space resulting in slow communications in locations well beyond Earth caused by the physical limits inherent in the speed of light and a low cadence of people coming in and out of a settlement could facilitate a St Kilda-like religious culture, possibly a cult, even alongside a fully functioning participatory democracy with modern communications technology.

Too much freedom as a form of extraterrestrial tyranny

There is a phenomenon that has been pointed out to me on more than one occasion in my travels to study communities on the Scottish Hebridean islands. It is not one that I could have guessed, or would have predicted through my own deduction, but once expressed, its reality is readily discerned. This phenomenon is too much freedom as a form of tyranny.

Living in an isolated, small settlement has its intensity of activity and demands with the many jobs and community obligations requested; but these are not unrelenting, and they sometimes leave substantial periods of quietude and calm which cannot be easily filled, particularly in places where there may be few of the appurtenances of towns and cities. The sensation about which I speak cannot be described as boredom; it is something different. It is the angst caused by confronting the open wilderness in which there are, at least outside of the immediate community demands, no obvious tram lines to guide one's life, a free expanse of time which is there to be filled; it is the unease of being delivered a form of pure freedom. The isolated settlement in extreme conditions seems to draw this emotion out with greater force than in most societies. In the literature, Alexander had said this of Colonsay:[50] 'The apparent freedom offered by island life could prove hard to handle for even those born to it, let alone those who were discovering it. Most islanders had a rare degree of control over how they spent their time.'

We can be certain that in extraterrestrial locations, settlers, particularly in early years when tasks will involve the upkeep of critical life support machines, will be busy, but, like Scottish islanders, this may not be uninterrupted, particularly if the automation and fault diagnosis of these systems is adequate. In stretches of downtime, staring out of a habitat window across a dead, dusty alien world, will they not suffer now and then, not from the autarchy of an authority, but from the tyranny of unfettered freedom? A tyranny of liberty.

The intensity of this state may be exacerbated by increasing automation. Driven by the lethality of the environment, there is a powerful encouragement to build life support systems that are reliable, have multiple layers of redundancy within them, and are not subject to the vagaries and unpredictabilities of human behaviour and error. In achieving such a technological state, humans will be liberated from the concern about death caused by the failure of these life-giving machines, yet concomitantly they will be delivered into the tyranny of having long periods of undemanding freedom within their physical confinement. The absorbing distraction of survival and its immediate needs will be replaced by the practical challenges of filling empty time in a physically limited space alongside the mental challenge of confronting human meaning and purposes that will bubble forth, involuntarily, in these stretches of quietude.

I do not think there is any particular antidote to this sensation, other than the opportunity to vanquish the feelings with things to do. These may include periods of conversation with others to fill time, or activities separate from work that provide a focus when freedom creeps into view. The proposed importance I place on voluntary association to dilute central control, which I discuss in the next chapter, will also see to it that there are plenty of activities to soften the trepidation brought on by an excess of liberty.

We should recognise that in the depths of space, a settlement has the capability to stir an experience of coercive lack of freedom and the dread of too much freedom all at once. Within this contradictory maelstrom and flux of liberties, there is a unique and disquieting form of human torment.

Extremities will channel tyranny into similar forms across the cosmos

My thesis so far has tended to treat all environments in space as similar with respect to their relationship to liberty, in that although I have discussed certain facets of locations such as the Moon or Mars, I have largely assumed the ideas presented here to be applicable to all of them.

It is the case that there are fundamental physical differences between various locations in space. On Mars, for instance, water is more easily secured from the atmosphere or from permafrost, whereas on the Moon there is no substantial atmosphere and there are no large areas of contiguous glaciers and permafrost. These differences may have an influence on the character of society and the liberties expressed. The extreme paucity of water on the Moon and thus the more difficult conditions to collect it may drive some of the conclusions I suggest to greater extremes than on Mars. In the expanses of free space, any new provision of oxygen and water must originate from a planetary body, and in these isolated extremes, we can expect recycling to become even more essential than on the Moon and Mars, perhaps with ancillary demands on people and their lifestyles, degrading some of the freedoms that are possible in less restricted social arrangements. In contrast, a two-dimensional surface to walk over on the Moon and Mars with various landforms and features may result in a liberated view of the capacity for freedom of movement compared to the homogeneous black emptiness of space (although one might observe that in a fundamental sense a two-dimensional surface is more restricted than a comparatively limitless three-dimensional space through which one might float). I think that we can expect every settlement beyond Earth to develop its own culture and for those cultures to be heavily biased, in some way, by particular planetary bodies and the accessibility of resources on them.

Yet, despite the obvious opportunity for some diversity, the hypothesis I wish to advance here is that the variety of conditions to be found beyond Earth is rather limited within the brute facts that oxygen is not readily breathable and in no extraterrestrial environments is there immediately drinkable water or naturally available food. All three of these conditions are combined in extremity and without reprieve everywhere. So the conditions under which people find themselves with respect to supplying the basic needs of life are rather similar, and we can expect that they would share a common view of these limitations. In other words, although the space environment admits of variety, the unyielding physical situation and isolation have a tendency to cause convergence.[51] Like the way in which biological convergent evolution leads to similar biological forms, these inescapable physical conditions cause a technological, social, political, and economic convergence in the conditions for society, with its implications for liberty. It is this underlying hypothesis that has led me to write generally of liberty in space, with little regard for the specific location.

It is intriguing to compare this to human history so far (Figure 2.3). As humans left northern Africa to spread across the globe, so they diversified in the environments they encountered. Their early hunter-gatherer lifestyles in the plains of Africa eventually gave rise to societies in rainforests, in temperate grasslands, on polar ices, at the edges of lakes, in deserts, on rain-draped mountaintops, in rocky caves. In all these locations, the sources of food and water varied, and although they all shared the common availability of free oxygen to breathe, the sheer diversity of geological, biological, and climatic conditions on Earth led to a veritable burgeoning of social types and arrangements. We might see human history in settled societies as a period of social blossoming, a colourful kaleidoscope of experiments in social organisation.

Yet, now, in expanding to an even wider vista in terms of physical extent—outer space—we are likely to find ourselves undergoing a restriction if not in the details of the cultures and ideas that might emerge in space settlements, then in the essential fabric of society mandated by the relative homogeneity of the space environment. That is not to say that we can envisage a unitary governmental structure that would be applicable to all places beyond Earth, or that we should try to create such a structure. But it is to suggest that when you have a group of people living under the potential of instant lethality and you must have systems of supply for oxygen, water, and food that can nowhere be easily obtained without great effort, the fundamentals of those societies will be remarkably similar.

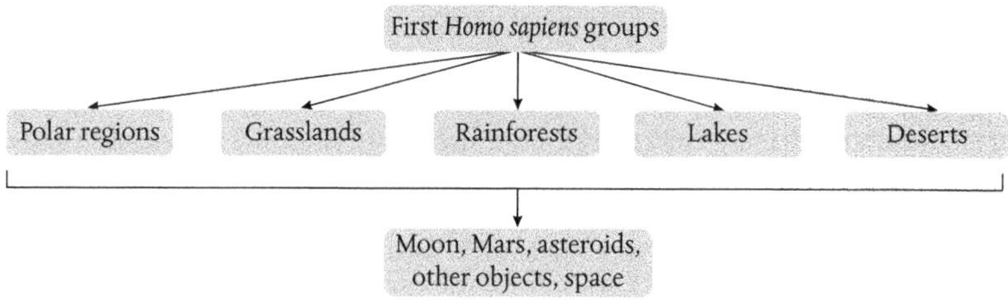

Figure 2.3. Social convergence beyond Earth. Humanity's breakout from North African plains has led to an efflorescence of political forms in a huge variety of geological, climatic, and biological conditions, with just some examples shown here. When we travel into the even more spatially extensive realms of space beyond Earth, however, the lack of oxygen, water, and food and the various social conditions this causes may lead to a convergence in political form despite the larger spatial area available.

To return to Antarctica, on this continent, every base has the stamp of its national culture imbued within it, but one is struck by the similarities between the bases that tend to hang overwhelmingly over these subtle cultural divergences. These similarities are felt in health and safety, functional equipment and paraphernalia to live in the polar extremities, logistics and travel equipment, schedules, and rotas, and so forth. The polar conditions clamber into every nook and cranny of life and cajole and force the total social infrastructure into a mould, like a jelly—you can change its colour, you can add flavours, lumps of chocolate, or coloured sugar pieces, but the dish within which it sets is unyielding and fixed. The extraterrestrial cast of the environments beyond Earth is the receptacle into which we will pour the collected variety of human cultures, ideas, and technologies. And as we dispense them, they will create swirls and eddies of variegated forms and shapes, yet all the time settling down into the self-similar conditions that define the most extreme environments into which humans have ever ventured.

I have so far identified some of the ways in which extraterrestrial conditions will channel the human social condition and organisation, focusing on those aspects that tend towards illiberal forms of governance. My next task is to discuss some of the overarching ways and means by which we can prevent these conditions from delivering extraterrestrial settlers into tyrannies.

3

BUILDING FREE SOCIETIES IN THE COSMOS

Our objective is to construct an anti-dystopia, not a utopia

It is attractive, and easy, to imagine utopias. If people were ready to act without flaws, then we could build societies with no conflict, where humans could live in blissful sympathy, erudition, compassion, and reasonableness.[1] Literature is replete with utopias, from Francis Bacon's *New Atlantis*[2] to James Harrington's *The Commonwealth of Oceana*.[3] In fairness to Bacon, his utopia was not an island of perfection as such: he was more interested in imagining an almost ideal science academy (at least in the part of his novel he finished), but his contribution is nonetheless an imagined state of unusual excellence.

We could easily imagine utopias in space where none of the concerns that I raised about tyranny exist, or ones where, if they do emerge, we resolve them by social arrangements never before witnessed in human society but likely to be successful beyond

Earth because the people will willingly submit to, and actively wish to construct, these new extraterrestrial Elysian fields.

Here, I avoid an attempt to imagine a utopian response to the matters I have discussed, not because anything utopian is by default indefeasibly impossible (although that is usually the problem with utopias), but because it is a poor way to allay the worst tendencies of humans. By assuming that people will behave as you wish them to, or in ideal ways that are motivated by their goodwill and self-driven desire to avoid bad outcomes, one opens the field to despotism precisely because one has not taken the care to seek out the conduits of poor conduct that lead to tyranny.

In an incisive rebuttal to a proposal by John Locke for creating legislation in Carolina in 1663, John Adams remarked: 'He should have first created a new species of beings to govern, before he instituted such a government.'[4] Although we should be ambitious in our visions of improving human conduct, it is the case that if we imagine forms of social organisation that ask too much of people, or do not align with the general features that we understand of the human character, then government is liable to either fail spectacularly or degrade into violence and despotism as the governed seek ever more extreme measures to make their schemes work. We should be realistic not only about human capacities, but also about the imperfect nature of the material we do have available.

My outlook of how to construct extraterrestrial societies finds strong sympathies in James Madison's view:[5]

> If men were angels, no government would be necessary. If angels were to govern men, neither external nor internal controls on government would be necessary. In framing a government which is to be administered by men over men, the great

> difficulty lies in this: you must first enable the government to control the governed; and the next place, oblige it to control itself.

This view of the fallibility of human character was a prominent and commonly held view of his revolutionary contemporaries who, because of the arbitrary injustices they perceived themselves to be subject to, were sensitised to the weaknesses of human power.[6] James Otis observed:[7] 'The unhappy revolutions which for ages have distressed the human race, have been all owing to the want of a little wisdom, common sense and integrity, in the administration of those, whom by their stations, God had in kindness to the world, rendered able to do a great deal, for the benefit of mankind, with the exertion of a small portion of private and public virtue.' In other words, it did not even require gross turpitude to render power malignant. Even a want of a little wisdom can make it tyrannical, and a society should be assembled on the basis of these human frailties.

Others outside the maelstrom of revolutionary America expressed similar sentiments. Catharine Macaulay, the English historian, had observed that 'all systematical writers on the side of freedom plan their forms and rules of government on the just grounds of the known corruption and wickedness of the human character'.[8] Furthermore, it takes little persuasion for these characteristics to become evident, given that the barriers to them were artificially erected by society and are therefore subject to degradation by the very human character that invented their antidotes.[9]

All of these writers were expressing a long-held view that humans are corruptible and that when offered the opportunity for power, it is not that people will *inevitably* take it and abuse it, but it is that there is no shortage of people who will do so, and a society makes itself vulnerable to such individuals when it offers unchecked power. As Machiavelli famously observed before them:

> As is demonstrated by all those who discuss civic life—and every history is filled with such examples—it is necessary for anyone who organises a republic and establishes laws in it to take for granted that all men are evil and that they will always act according to the wickedness of their nature whenever they have the opportunity, and when any wickedness remains hidden for a time, it arises from a hidden cause that is not recognised by those who lack experience of its contrary, but time, which people say is the father of every truth, will eventually uncover it.[10]

This is sound caution, and I have sometimes thought that Machiavelli's Machiavellian tendencies were somewhat overstated. Some time after the American revolution had achieved its objectives, political philosophers continued to regard this problem as a central precept, or foundation if you will, of the construction of good government. James Mill, in the nineteenth century, for example, averred that:[11]

> This has ever been the great problem of Government. The powers of Government are of necessity placed in some hands; they who are intrusted with them have infinite temptations to abuse them, and will never cease abusing them, if they are not prevented. How are they to be prevented? The people must appoint watchmen. But quis custodiet ipsos custodes? Who are to watch the watchmen? The people themselves. There is no other resource; and without this ultimate safeguard, the ruling Few will be for ever the scourge and oppression of the subject Many.

The human tendency to tyrannise is a statistical problem. There are many good individuals with power who act with probity and trustworthiness, although they are apt to assume that others will, or should, behave like them. However, human honesty is distributed in some function that covers the full range: from near-angels to veritable rapscallions and all points in between. When one is presented with a sufficient population, one can be assured that there will exist within that body of people those who represent the worst end-member examples of that distribution. This is the essential statistical matter that these political philosophers and others recognised, and that we must comprehend for the extraterrestrial case.

In essence, what we wish to create in space (and I would contend on Earth as well) is not a utopia, but an anti-dystopia. We seek to identify the holes through which the freedoms of people leak and strengthen and cover those weaknesses before they should be the conduits through which liberties will drain. These attempts themselves can never be perfect, but they may have the possibility of creating societies that do not descend into tyrannical dystopias.

It might be useful to try to define at the highest level what it is that we wish to achieve. Of all the political philosophy that seeks succinct definitions of the aims of a free society, I find Friedman's brief statement to be emblematic of what I think we are trying to do, or at least an accurate summary of the underlying ambition in my own thoughts on extraterrestrial liberty:[12]

> These then are the basic roles of government in a free society: to provide a means whereby we can modify the rules, to mediate differences among us on the meaning of the rules, and to enforce compliance with the rules on the part of those few who would otherwise not play the game. The need for government in these respects arises because absolute freedom is impossible. However attractive anarchy may be as a philosophy, it is not feasible in a world of imperfect men.

In essence, we seek to achieve the balance between collectivism and individualism by ensuring that on the collectivism side of the equation, the rules that establish the obligations we have towards one another are decided by mutual consent through continuous deliberative processes.

Before I progress to consider some social structures that could be used to thwart tyrannous tendencies in space—to realise the political structures of an anti-dystopia—I first wish to address a potential misunderstanding about the view of the individual in society and the apparent antagonisms between individuals and society that I drew attention to in the previous chapter. This is essential for understanding the notion of power in space and the relationship of the individual to it.

Centralised power is an inevitable feature of human existence in space

The way in which I have described the potential for tyranny to emerge in space could be misinterpreted as a view that power structures are always unfavourable and that the individual, as an isolatable entity, is to rail against power to protect their limited sphere of rights and freedoms. In this view, central power is an unwanted artefact of society that individuals must fight against. This is a fundamental misunderstanding. Wherever individuals collect, and particularly in space, where the conditions are so extreme and require coordinated action to survive within, there will emerge centres of power.[13] Indeed, these centres of power, even if they are small participatory democratic institutions, are desirable and necessary to shape the collective development of society in extremity. In short, in this work I set the individual as under threat from power because this is the most parsimonious way to identify all the places from which tyranny might come and then consider ways to block them, but it should not be misunderstood as a statement that the individual stands completely aloof from power and should, in some ideal society, be utterly protected, as an isolated creature, from it.

The misperception of the individual as an atomised entity trying to defend him or herself from these 'outside' power structures is encouraged by the wider misunderstanding that a liberal view entails precisely this idea—that individuals precede society and all social structures must be constructed with this highly individualised human being in mind.[14] However, this, I think, is also a caricature of the liberal mindset. Even classical liberal philosophers of the Enlightenment who were keen on using the device of humans in the 'state of nature' to consider how people agree to come together to assemble societies were careful to point out that they never believed in the reality of human individuals alone and separated who came together to work out a compact for their cooperation (although perhaps Rousseau believed this in some literal way). The state of nature was a thought experiment, a device to consider how human organisation should work.[15]

Stemming from this—the notion that classical liberal ideas are naïve because they attempt to place the individual prior to society and that we can aim for individualism that stands in isolation from these wider power games—is, I think, a misapprehension. I have little hesitation in saying that some libertarians might actually hold this outlook, but classical liberalism as such never saw the individual as an atomised entity. Adam Smith, the archetypal classical liberalist, committed an entire book to exploring the 'sympathy' and moralities that define the collective conditions of society and the relationships that emerge between people.[16] These notions are of relevance to the extraterrestrial case, where a society will be constructed from scratch. There may be a temptation to consider such a group of people as representing a literal manifestation of a state of nature in which we must consider a social contract to bind these separate entities together into an ideal polity. Yet, the acceptance that power centres and structures will exist from the outset is necessary to shape them satisfactorily.

The existence of centres of power and their influence over the individual is a fact of life for a group of social creatures; dominance displays in other primates are an assertion of hierarchies and existed long before human societies. Settled human societies emerged from their inception as collectives with hierarchies and unspoken 'contracts' of power between individuals.[17] Furthermore, many liberal thinkers have themselves pointed out the way in which powerful entities in society, by being in competition with each other, may inadvertently protect the individual from usurpation by these entities, in these cases demonstrating the importance of power structures and their competition in allowing for the emergence of individual freedoms in the first place.[18]

It is also trivially the case that where centres of power exist, these centres and their component people tend to use intermediate centres of power and individuals to get what they want and that many claims to individual rights can result from powerful groups using those individuals to achieve ends, reducing 'individualism' to the interplay of power centres in society.[19] However, this phenomenon is precisely the reason for concepts such as the rule of law, equality before the law, and freedom of expression. If centres of power were not a fact of life, these ideas would be redundant; individuals would live their anodyne and happy lives devoid of the need to give expression to such aspirations. The very use of the word 'law' in some of these dictums is an explicit recognition of the existence of legislatures, judiciaries, and other systems of authority. They are an acceptance, not a repudiation, of the power relationships within society; they are an attempt to canalise behaviours that result from power politics in constructive ways that reduce the arbitrariness with which individuals get caught up within them. They recognise that wherever collectives of humans exist, individuals can be mistreated because of their often subordinate power status. Therefore, the concepts just elaborated, and others, are an attempt to establish constructive and positive rules between individuals and these larger structures of power. In that way, a liberal view embraces the reality of collective power but seeks to control and shape it in the interests of maintaining a domain of individual freedom.

Another complication that causes much confusion is that there exists a modern use of the word 'liberal' that might be said to encompass this more singular view of the individual seeking his or her rights backed up by a system of state power, with the individual viewed as essentially prior to society and society's purpose as catering to the individual. Now whether one thinks this modern version of liberalism is desirable is not a matter I want to address here. This is not a general book on the history of liberalism. The only point relevant to this work is that I wish to emphasise that the often antagonistic view of power that I express in this work and my view of the individual being protected against it should not be misconstrued as reflecting this perhaps more modern atomised view of people. By focusing on where friction might lie between power structures and a sphere of private individual activity, my goal is to ascertain some of the ways in which the relationship between the two might be improved and coercion might be avoided. The relevant matter is that the extremities of space, in certain spheres of human activity, will enhance the excuses for tyranny; therefore, we should fashion societies and infrastructure in such a way that individuals and these larger centres of power can work productively in unison and the individual is allowed some semblance of private activity.

A more collectivist form of liberty is a likely requirement in space

In devising ways to propagate liberty in space, we might take our current perceptions of freedom (or to be more exact, my current Western perception of freedom), expect it to be replicated in space, and then attempt to imagine all manner of mechanisms to push forward this vision of liberty, using ever more extreme methods to make it work, even if it apparently 'won't fit' in the extraterrestrial environment; it would be easy to doggedly attempt to shape our own ideals of liberty into these alien conditions. However, this seems to me to be both futile and fallacious.

At some level, we have to accept that the space environment creates a high degree of interdependency between people, and that the relative difficulty, compared to most places on Earth, in acquiring basic commodities, fused with the natural restrictions in movement, will curtail freedoms. Might we not accept that a more collective social culture would emerge? I think that the straightforward answer to this is in the affirmative, and if we want a historical perspective on how this liberty might be different from what some of us experience on Earth today, we might look backwards, not forwards in time.

Benjamin Constant famously compared the liberty of the ancients to the moderns, a comparison I have previously applied to the extraterrestrial environment.[20] Constant's fundamental point was that in ancient societies—take ancient Athens, but it could be other city states and collectives—the lack of coordinated social infrastructure, machines, and so forth, coupled with generally small populations compared to most modern nation states, created societies where the interdependence required to survive meant that it was more necessary to integrate oneself into a collective than it is for some of us today.[21] Put simply, it was less easy for people to live the highly individualised lives that some humans can fashion in the present day on account of labour-saving devices, public transport, and the milieu of combined efforts of many millions of people that make it more easy to carve out such a life. That is not to say that we are no longer social animals; even if you live on the edges of society, you are probably still dependent on machines, electricity, processed clean water, and so forth, which are very much the products of the collective effort. It is just that you do not have to be aware of this background of assistance on a day-to-day basis. Before the reader should castigate me, I accept that these observations do not apply to a vast number of people on Earth who do not enjoy the blessings of a technologically organised liberal society that allows for the expression of individualism in the way I have presented it. But like Constant, I am trying to make a comparison to an experience that many readers of this book will be familiar with. Constant's essential point, I think, is valid, although we might argue about the specifics. The ancients had a more collective view of society than those of us in the West have today.

In the extraterrestrial case, the general goal is to avert extreme forms of despotism that leave people enslaved, but we might accept that the societies that exist are likely to embody a higher degree of social collectivism than some of us are used to. Constant's liberty of the ancients becomes the liberty of the future. Nestled between the past and the future are the resource-rich expansive nation states of Earth enjoying the benign living conditions of the planet, which allow forms of detachment and individualism that to the ancients, and perhaps also to a person tending an oxygen machine in the dusty desiccated expanses of Mars, may have seemed perverse. I do not know whether this vision of liberty is correct, but the only point I wish to draw from it is that in attempting to create societies in space where liberty is on the minds of its occupants, our desideratum is not to go to extreme lengths to create maximum liberty as such, where those interventions necessary to achieve this become a form of coercion in themselves, but to find a balance wherein the interventions are sufficient to prevent a slide into abject tyranny. This balance may engender a more collective view of social behaviour and interaction than some more liberal twenty-first-century terrestrial conceptions of liberty might admit.

Further, in accepting a more collective view of space societies, we should not make the mistake of seeing this as a compromise, a step back from an ideal. As one of Constant's 'ancients', Aristotle was keen to point out how the individual found their meaning and power in being part of the city state.[22] Humans are, as Aristotle would have it, *zoon politikon*. The polis provided the context in which these 'political animals' achieved the expression of their capabilities, which otherwise were isolated and relatively meaningless.[23] As Pericles had said, 'We consider anyone who does not share in the life of the citizen not as minding his own business, but as useless.'[24] The person who tends an oxygen-producing machine on the Moon is finding within this machine a sink for their skills as an individual; by contributing their personal capacity to its upkeep, they contribute to the society as a whole. Now although this observation may seem facile—one could say this of almost any job on Earth—it is certainly the case that in Western societies, as I have said, we tend to be less emotionally aware of this idea that the expression of our individualism occurs by being part of a collective. It may also be true to say that we are more wary of this idea as the thin end of a wedge to all sorts of excuses for control that might emerge from pushing this argument into justifications for imposing draconian collectivist policies because 'this policy of social control will help you achieve your individualism'. This sort of cynical misuse of the relation of the individual to society brings back foreboding memories of Rousseau's[25] notion of persons being 'forced to be free' to align them with a 'general will'.

The truth of this intersection of individuality and collectivism in a way that might seem quite contradictory to many modern minds can be found in the Israeli kibbutzim, a small-scale socialistic commune. Spiro, in examining the sociology of these communities, observed:[26] 'Another principle underlying the culture of Kiryat Yedidim is that of individual liberty; indeed, the kibbutz prides itself on being the freest society in the world.' This proclamation seems in contradiction with the following statement just one page later: 'It means first, that the interests of the individual must be subordinate to the interests of the group.' If one understands that you can only be an individual in such an extreme environment by belonging to a group, then these two observations are not seen to be in contradiction by those who pronounce them. This apparently conflicting view of individuality is an immensely important one to grasp with respect to extraterrestrial conditions. Furthermore, it needs no overt submission to a socialist vision, as in the kibbutzim, to realise this outlook; the extremity of the extraterrestrial conditions and the need for coordinated action may well achieve that regardless.

It may be the case that in relinquishing control to a collective, an individual does feel liberated in a very real sense. This *can* have a positive lining in facilitating a sense of free expression achieved by making a readily identifiable contribution to society, by being part of a larger whole; in the right measure, it is a laudable sense of our humanity. However, we should be mindful of the darker side to these impulses that drive this positive sense of community to the other extreme. The sense of liberty animated by submitting completely to an all-powerful governance can lead inexorably to a totalitarian condition in which individuality is extinguished.[27] Free of the need to make decisions about their own lives, and given the capacity to blame any failures in life on 'the system', over which they could have had no control anyway, the abnegation of individuality can become a form of liberty: escape from the responsibility and confusion of freedom is a type of liberation. But there can be no rational human being who would demure to this outlook and simply declare that accepting this state of affairs is just another form of liberty, as valid a manifestation of freedom as any other. In embracing this form of liberty, not only does the individual destroy all other forms of self-expression, but they supplicate to a social order wherein all alternative types of freedom, civil, political and economic, are likely to be utterly destroyed. The extraterrestrial society should guard against the acceptance of a more collectivist mentality collapsing into this form of submission. I will have more to say shortly about how we might avert this end.

Despite the modern reticence we may have in linking the realisation of our individual capacities to a society that makes strong demands on collective behaviour, a reservation that is healthy since it encourages a constant vigilance, in extraterrestrial environments, where an individual will depend more comprehensively on others for survival, this general notion of collectively realised individualism will be more imminently felt.

In practical terms, therefore, we should embrace the extraterrestrial polis and realise that the required efforts, capacities, and demands in surviving and thriving in the extraterrestrial environment can nurture the individual and allow the blossoming of their abilities. A return to Constant's liberty of the ancients is not a retrograde step, but a pragmatic realisation of the needs and wants of a society living at the edge of physical limits.

Constitutions will not be enough to secure liberty

We have established some of the general culture of the society we wish to create. In beginning a journey into the practical means by which we might establish liberties beyond Earth, the first and most likely response will be that we merely need to create constitutions that enshrine rights and formalise a free system of government.

When Alexis de Tocqueville toured the young United States of America, making observations on its people and system of government as he went, I think that one of the most pertinent observations he made with application to the extraterrestrial case concerns constitutions. He observed that despite the similar geography and environment of Mexico, and in spite of the efforts to copy the format of the US Constitution, Mexican society had not created anything like the attitudes towards liberty that had taken root in the US.[28] Why was this the case?

His answer was, to precis his observations, simply that the people of the US had, as they sailed to the New World, brought with them more than their supplies and victuals. They had emigrated with an attitude: puritanism, a spirit of hard work, and a desire for religious and personal freedom. These outlooks were embedded in their view of the world, and with them came personalities that were tired of tyranny and desirous of freedom. Tocqueville's point was that constitutions were not enough: a culture and desire for liberty were the essential pre-requisites. Customs were as much a part of free societies as constitutions. Constitutions gave these proclivities form in the written word, but they were not sufficient. The essence of what he saw, and elaborated in detail, was observed by John Adams before him, who, in commenting on Mary Wollstonecraft's *French Revolution* and her observation that a constitution is a pillar of government, had remarked: 'A constitution is a standard, a pillar and a bond when it is understood, approved and beloved. But without this intelligence and attachment, it might as well be a kite or balloon, flying in the air.'[29]

Much later, views similar to Adams were expressed by Judge Learned Hand in a speech he made (*The Spirit of Liberty*) in 1944 in celebration of *I Am An American Day*: 'Liberty lies in the hearts of men and women; when it dies there, no constitution, no law, no court can even do much to help it. While it lies there it needs no constitution, no law, no court to save it.'[30] What Adams, Tocqueville, and Hand recognised was that constitutions—formally identified rights, even—are not the guarantors of liberties, and their mere presence will not ensure that individuals can avail themselves of those rights, or that they will be respected. Even when the authorities do respect defined rights, a constitution will not ensure that the people themselves have the temperament to cherish and uphold them. The essential matter, which many have recognised, is that they are necessary but not sufficient.[31] The most masterful exposition of the boundaries of liberty and its protection in the most exquisitely crafted document cannot alone create a free society if the mores and morals of a society do not support it.[32] In contrast, the inverse conclusion should not be drawn: that constitutions are valueless. A constitution can formalise the ideas of freedom and provide a reference point that can be consulted when those freedoms seem to be abridged or threatened.

It is probably not difficult to see why Tocqueville's observation stands. Even before he proffered his ideas, the 'constitution' of a country was thought to enclose much more than a written document; indeed, the almost sacrosanct nature of the written US Constitution, which gave force to the idea that such a document could be the supreme governing document, not only a useful elaboration of some powers and rights, was a relatively new concept of governance. Charles Inglis, in 1776, had spoken of the constitution as 'That assemblage of laws, customs, and institutions which formed the general system; according to which the several powers of the state are distributed, and their respective rights are secured to the different members of the community.'[33] Thus, during the American revolutionary years, some people saw the general constitution of a nation to be much more than whatever could ultimately be expounded in a written form of constitution, a word that would ultimately be capitalised when manifested as a supreme governing charter; Inglis had specified the importance of customs.

Beyond Earth, much of that unwritten bastion of liberty will come from both history (in other words, the culture and ethos of freedoms that the settlers take with them) and that ineffable quality of 'virtue' so beloved of the Roman Republic, so easily treated with disdain in the modern world, yet so crucial to ensure that the written constitution, and the unwritten rules and manners that surround it, are respected. History has shown, as Tocqueville saw, that if there is a potential weakness for the founding of an extraterrestrial free settlement, it will be in these qualities that fail to carry the day for liberty, despite the best efforts embodied in political design. It might behove those who select the first settlers in any location in the Solar System to consider the required character traits beyond merely the functional questions of whether the people they have selected are capable of the physical and mental exertions required for survival. They might also probe whether the proposed settlers have an understanding of political and constitutional orders, a love of liberty, a deep knowledge and some scholarship in the literature of liberty and political philosophy, a proven virtuous character that eschews power for public duty and service, and a mind that sees the end state of social harmony and freedoms and that can weigh those ends against the short-term passions, jealousies, and rivalries of human affairs. These are the characteristics that one should seek in abundance in a founder population. They are the rivets in the grander structure of the constitutional ship, a form that tends to attract the attention of the world, yet cannot be held together without those vital unseen links.

By selecting a promising starting population, the initially stochastic, unpredictable historical conditions in which the founders find themselves may be less likely to cause liberty to founder. Even with the best intentions, culture, design, and constitutional documents, there are still the random fluctuations of circumstances that can never be planned for, yet whose negative consequences could be mitigated by people with suitable qualities. In considering the factors that shaped Pennsylvania, and William Penn's involvement in its founding, Andrew Murphy observed: 'Studying Penn and Pennsylvania suggests that even when the founding in question is overwhelmingly devolved onto one individual, with disproportionate influence over process and substance, founding intentions are often still indeterminate, perhaps inevitably so, insofar as colonisation is always a collective endeavour, never fully controlled by a single individual or even one group of individuals. Founding moments serve an important purpose by highlighting the significance of factors such as contingency, interest, material considerations, and/or conflicting goals alongside the important role played by theories and written constitutions in the building and maintaining of governing structures in often-precarious conditions.'[34]

For the settlers of the new worlds beyond Earth, written constitutions will be invaluable for circumscribing the limits of central authority and stating rights, such as the freedom to leave or the right to oxygen, which they might demand from their leaders. In this work I will point out matters that might merit constitutional recognition in extraterrestrial documents of this sort; these documents may also lay out the basic structures of governance and the means by which those structures are ordered. But above and beyond these charters, these societies must use many other mechanisms to foster ideas of liberty in the minds of their people.[35] These other means are the subject of much of this work.

Constitutions should be a matter for a settled population who will abide by the constitution, not for those outside the settlement, including those who are bound to Earth

Notwithstanding the points I have made about constitutions, the formulation of written constitutions will be desirable as one element in the assembly of a free republic. Is it not the case, then, that we should set about this task now to provide these future settlers with at last one ready-made instrument in the tool shed necessary to build such a society?

There is an element of entertainment and the appeal of a long-term legacy to devise extraterrestrial constitutions for future lunar and Martian societies. After all, as we have seen, the extreme conditions are predictable, and we now know enough about many extraterrestrial environments to be able to contemplate the challenges that societies might face and how they might be overcome.

In the next section, I will touch on particular rights that may be incorporated into a constitutional document, but I think there is considerable hubris in putting those ideas together into a formalised constitution for a people who do not exist or for a society yet to be constructed. Ultimately, it is those who must endure the extraterrestrial environment and the political situation it brings forth who must write these documents. As the identification of rights, more than merely the absence of unwanted domination and interference from authorities, constitutes part of the structure of civil liberties, it is necessary that such rights are debated, identified, and then prioritised by those who will benefit from their existence.

It seems to me to be a matter of weighty principle that constitutions should be devised by the people who will live under them. If the British had attempted to draw up and propose a constitution for a young America and then offered this to the revolutionaries who sought independence, I have little doubt that it would have seemed to contravene the very notion of independence and would have been torn up. How much more absurd is it to imagine that we should be any less arrogant to devise constitutions for future societies in space that have yet to come into being? A Martian Constitution should be designed and written by Martians.

This view need not be taken to extreme moral degrees; there is nothing obnoxious about a group of people on Earth dreaming up a Martian governing document for the fun of it, as an educational exercise, or as an interesting intellectual task, for instance.[36] However, when the time comes, the consolidation and coalescence of principles and rights required for a real constitution must be carried out by the people who are embedded in the physical, political, economic, and social reality of a new extraterrestrial society. They should consider themselves to have free rein to devise an entirely new document that owes nothing to previous attempts at such instruments and nor should those people, if they are still alive, who attempted to construct such constitutions outside these societies, feel any sense of chagrin that their ideas are entirely ignored.

To desist from drawing up extraterrestrial constitutions and honour the principles of liberty by deliberately restraining oneself from such activities would be a mark of civility towards future space settlers and a form of respect, deference even, to their independent minds. For this reason, there is no attempt in this work to devise a constitution, although I will set out the forms of liberty that could ultimately be protected by such an instrument.

On extraterrestrial rights and the use of familiar rights in space

Although we might rightfully shy from the construction of constitutions, I do not think that it is presumptuous to consider some of the rights that extraterrestrial settlers might expect and that ultimately could be enshrined in such documents. The liberties to which I have given most attention are the sorts of civil and individual liberties one achieves by controlling and subduing domination, interference, and arbitrary political power. However, an important way to shore up some of these freedoms is to define specific rights that could be stated in governing documents and embodied within the legal and cultural milieu of the settlement.

When the United States Bill of Rights was conceived in the eighteenth century to appease disquiet about the power of the newly devised federal government, one claim for its superfluousness was that any rights not explicitly delegated to the federal government did not need to be stated, since they naturally inhered in the people. Ultimately, the desire to elaborate specific rights prevailed. Likewise, in the extraterrestrial case, we could observe that there are certain characteristics about that environment, which I have discussed at much length already, that strongly recommend the affirmation of specific rights rather than the mere assumption that if they are not formally delineated in governing powers, the authorities will have no reason to infringe them. On the other hand, it would also be reasonable to assume that any right not explicitly adumbrated in a governing document should be presumed to be reserved to the people and such a credo might be stated directly, as it is in the US Constitution.

Some of the explicit rights that could be incorporated into any extraterrestrial charter or constitution are consistent with matters I have discussed already, and they include all of those rights that maximise the capacity of free expression in the extraterrestrial environment against that conformity and tendency to repression that I have identified. They include the right to religious freedom, the freedom of the press, freedom of speech, the freedom to assemble, and the right to seek redress of grievances from the governing structures.[37] The US Bill of Rights achieves all of these in a single constitutional amendment (its first amendment[38]) and this simplicity seems eminently transferable to the extraterrestrial case. As with terrestrial rights, these are not absolute; rights are associated with correlative duties. Freedom of speech does not imply that I can stir up public disorder on the Moon and then cajole all around me to cause a serious depressurisation event without consequence. The protection of rights implies a parallel form of responsibility from citizens in using those rights. On this matter of the virtue of the citizen body, I will have more to say later.

Related to free expression is the capacity to feel secure in one's habitat and personal space. Except in extraordinarily severe exigent circumstances, such as failure of an oxygen system, the governing powers should be prevented from forcible searches and violations of these spaces unless they have a good cause whose exact purpose is stated and is subject to public view, such as through electronic media. This right is not merely expressed through non-interference; it entails all of the institutional structures that reduce the coercive behaviours of the governing bodies that can lead to its violation. If extraterrestrial settlers are to have the capacity to express their own views without fear, they must not be at the mercy of the presumed benevolence of those who run the settlement. This right, alongside those governing how a person is to be treated in matters of law, is essentially a means to guarantee the right to personal security. This right is concordant with the fourth amendment in the US Bill of Rights[39] and might also be easily transferred to the extraterrestrial case with little modification.

In case the reader thinks that I am lazily relying on repeating some of the amendments of the US Constitution, let me observe that much effort and thought by the founding fathers went into devising these rights, and it seems superfluous to start from scratch because of a squeamishness in blandly repeating them, provided, of course, that we think about them and decide to agree with them. Here we have an excellent example of how the lessons of history and the long trains of thought that have been invested in political philosophy on Earth can provide us with a basis for assembling societies beyond Earth.

It is apposite to draw attention at this point to the fact that although the language of rights used here may seem to be the domain of nation states, there is no reason why the principles I am discussing would not be applied in the smallest settlements. A governing document of a lunar station of just a few occupants could state the rule that an individual's private quarters will not be subject to search by others or the station commander. Thus, the points under deliberation here apply to all scales of human settlement, although we could expect them to take on a more formal visage as rights in more elaborate constitutions as a settlement grows.

It is necessary to identify rights, which could be amalgamated into a single stated right, that relate to the operation of the judiciary. I will discuss criminality in a separate chapter, but here I mention that they are the right to an impartial trial, to hear the reasons for the criminal charges, to use and confront witnesses, to be protected from double jeopardy and self-incrimination, to be treated humanely, and to have the right to due process. In small settlements, where the judiciary may be little more than a single individual enforcing rules, these rights may seem unnecessarily grandiloquent, but on inspection, each of them apply at any scale to the way in which an individual is accused, how they are treated, and the manner in which evidence is gathered. The lofty language of rights may have a salutary effect, raising awareness of the seriousness of criminal accusations, even in the smallest of communities.

On the creation of new rights in space and the problem of multiple extremities in the identification of those rights

Beyond these familiar rights, which can seamlessly uplift themselves from Earth and take root in alien places, there are specific conditions in the extraterrestrial environment that may merit more original statements. Later I will discuss the need to ensure a basic provision of water, food, power, and, in particular, oxygen (to avoid the appalling vision of individuals being subject to coercion by the threat or real denial of this vital gas); rights to these provisions, to ensure that no person shall have oxygen used as a means for their arbitrary abuse, might be enshrined.[40]

Let us consider the provision of oxygen. Intriguingly, Paley[41] is one of the few early political philosophers who made a statement of direct relevance to the extraterrestrial oxygen problem. He wrote: 'Natural rights are, a man's right to his life, limbs, and liberty; his right to the produce of his personal labour; to the use, in common with others, of *air*, light, water' (my italics). Paley was not referring to access to oxygen. Most likely he was referring to air that was not industrially polluted. Nevertheless, he inadvertently constructed a statement that could easily be incorporated, with modification to a more modern lexicon, into an extraterrestrial constitution to establish a basic right to oxygen.

In such rights, we are not merely expressing a desire that people have some access to a basic level of subsistence, but stating that access to such commodities is fundamental and associated with correlative duties from the settlement authorities to ensure that they can be met. Such rights might manifest as requirements for those who oversee a settlement to ensure the capacity to produce some minimal level of oxygen as required from the settlers.

The right to oxygen is not without contention. Many of the rights expressed so far, such as the right to free expression and freedom from arbitrary arrest, are largely negative rights in that they involve non-interference by a governing authority. The right to oxygen is the assertion of a positive right to some commodity, with its implications for others to meet that demand. The assertion of such a right is evidently open to autocratic abuse in that an authority can benevolently claim to be realising this right through the redistribution of oxygen production, which could involve interference in private associations, forced movement of people, interference in engineering within private spaces, and other apparently necessary actions to ensure the equitable distribution of the gas to those who are at risk of going short. These types of possibilities have always been the central matter of contention with proposed rights that have these more 'positive' implications.

One could argue that if a society was to hold that the right to life and liberty was enshrined in its constitutional order, then by default any denial of oxygen by an authority would be a denial of life, so in some sense the right to oxygen and the prevention of its abuse as an instrument to threaten the lives of the population could be satisfactorily encapsulated by other higher-level statements of social order. Nevertheless, an explicit statement of a right to oxygen might ensure that the possibility of the inadvertent lack of oxygen, through no direct fault of the governing authorities, would place demands on them to be sufficiently alert to oxygen production to ensure that basic demands could always be met. Whereas one could argue the same for water and food, it is the immediacy of death caused by the lack of production of oxygen that might warrant its elevation as a specific right.[42]

Are there other extraterrestrial rights with similar implications of positive duties from the authorities? Accepting the enormous potentialities for political and economic coercion caused by extraterrestrial isolation and the denial by authorities of a route of exit, the settlers might have the right to leave the settlement established as fundamental. James Schwartz has elaborated on the rationale and philosophical basis of such a right in the extraterrestrial case.[43] Its provision may do much to weaken a sense of helplessness if an individual feels held under duress in a settlement where they are suffering alienation of any kind.

The matter of which social factors in the extraterrestrial environment should be the subject of rights or instead established as desirable endpoints that might be established in rules, regulations, and laws is one that merits a great deal more discussion.[44] As I have pointed out on the matter of constitutions, these discussions should be undertaken by settlers and decided by them. For example, consider the matter of highly toxic dust: should it be a right to live in conditions where potentially toxic dust is not a threat, placing a demand on an extraterrestrial authority to put in place the regulations and rules to prevent such material from entering into living spaces? Perhaps on the Moon, where this problem is both serious and pervasive, and on Mars, where the dust contains potentially toxic perchlorate salts, we might elevate this to a right, to ensure that it becomes a priority action by the authorities to protect their people. However, in the free plateau of space, where there is no significant amount of dust, clearly this threat is irrelevant; we can see that there is variability in the extremity of certain problems that may come to the attention of those seeking to define rights, depending on where they are living. The pervasiveness of these problems will be influenced by the technological solutions that exist and the experiences that settlers have in coping with them, which are difficult to predict a priori. Thus, even if we were to encapsulate the right to oxygen within a broader category of rights securing a healthy environment, the components of such a healthy environment, such as the dust hazard, would clearly vary depending on the location.

Given the variation in conditions between extraterrestrial locations and the different ways in which they might be used in abuses of power, there is clearly much debate to be had among extraterrestrial settlers. Although some of these rights, such as the right to oxygen or to leave, may emerge as universal provisions, we might expect charters and constitutional documents to vary depending on the outcomes of these deliberations in any given society.

A secondary consideration is the extent to which certain of these living standards need to be elevated to a right at all. For example, the pervasive presence of ionising radiation in many locations in space leads to a need for shielding to achieve levels of radiation that are within some defined cumulative limits acceptable for human health. In a permanent settlement, should it be a right to be exposed to some agreed maximum level, or instead, might it be less formally inculcated as a regulation or a law pertaining to industrial standards for habitats and pressurised buildings and vehicles? For some human needs, we can see that laws and their associated regulations might do some of the heavy lifting before we make demands through rights.

The challenge of defining rights in space and what the practical correlative duties would be is that there are an enormous variety of items and products with which a lack could very rapidly yield indigent or even potentially fatal conditions; in this section alone, I have discussed oxygen, water, food, power, toxic dust, and radiation. There is the matter of healthcare, and later I will suggest that a minimum healthcare capacity should be assured. But what about clothes in the extraterrestrial environment? What do we do about a minimum level of entertainment in the lethal and potentially mind-destroying boredom of confined environments in space? Do we have a duty to provide these things, and do the settlers have a fundamental right to them? It seems that without them, the settlers' capacity to realise even the most basic of liberties would be severely curtailed. As a consequence, we could start to list a large number of items which, if they are not provided, could dramatically alter the human condition and render talk of life and liberties meaningless. Which of these, if any, are to be demanded under the banner of a right? And how do we ensure that the rights provided to the settler do not become so extensive as to cause society to degenerate into little more than a group of confined people constantly making constitutionally and legally mandated antagonistic demands on their authorities in their state of utter dependence?

I think that some of the answer to this distinctly extraterrestrial problem lies in the relentless pursuit of pluralisation in the means of the production so that a surfeit of items, of as many kinds as possible, is realised.[45] Under such conditions, the provision of clothes, for instance, need not become a right, but something for which the demand will encourage automatic productivity in the economic sphere. However, for a requirement such as oxygen, for which short-term want by any individual could be instantaneously lethal, and thus a conduit for coercion, we see strong reasons to elevate this to a right to ward off this outcome.

The concluding point I want to make is to recognise that a significant outcome of freedom of speech beyond Earth in the early stages of permanent human settlement will be the agreement of these very relationships between rules, laws and rights. Perhaps only those people living long term under extraterrestrial conditions will have a fully formed intuitive sense of the balance between rules, laws and rights, particularly where those proposed rights have more positive demands inhering within them. It is to the general methods of securing liberty and encouraging these discussions that I now wish to turn.

Democratic systems of government are optimal beyond Earth

The expectations, rights, and responsibilities of citizens are all, of course, fashioned by the wider political environment in which people find themselves. Perhaps the most complex arena in which we can discuss the construction of a free society in space is in this political sphere. It is complex because we do not empirically know what works and what does not, because we have not yet constructed a society in space, and because the varied and expansive number of possible mechanisms and details that one might design and develop to build a free society are legion. Nevertheless, despite these complications, we might attempt some comments about the general structure of government in space based on the observations I have made about the conditions for despotism.[46] In this section, I wish to discuss some characteristics of governing power in space before examining some particular suggestions in the following sections. I make the assumptions here that it is possible for a governing power in space to achieve legitimacy of some kind and that the space environment does not pose an unusual insuperable barrier to the legitimacy of a centralised power structure.

I have previously explored the legitimacy of something akin to an extraterrestrial 'state' in the context of traditional arguments about state legitimacy.[47] In essence, a contract theory of the state must be credible beyond Earth, because there is no possibility of an individual existing in a genuine state of nature. In simply being in a place where the oxygen they require for every second of their existence is provided by technology, they imply tacit, if not explicit, consent to the infrastructure that surrounds them and the people who keep it going. A utilitarian view of a governing power is somewhat specious, since not only does a governing power make people 'happier' in a utilitarian sense, it also allows them to exist in the first place, whatever its form. Other arguments for legitimacy tend to reach similar sorts of conclusions for extraterrestrial extremes, where some sort of highly organised order is required for people to be alive. In that sense, the most egregious forms of governance can easily claim legitimacy. From this point of view, then, arguments about the legitimacy of power are not very informative; the question is rather which forms of a priori legitimate governance are best for human flourishing and are not tyrannical.

It is likely the case that there will be some sort of centralised authority in space, not only because power centres naturally emerge in any human society but because, regardless of the inevitability of such power, it is actually desirable.[48] Faced with a depressurisation event, a lack of water, a serious fire, or other such critical situation that may become an existential risk to the entire settlement, a type of central coordination is necessary. If some group of people is to have sufficient knowledge to make coordinated decisions and action on behalf of the settlement, then it implies the existence of a central authority.[49]

We could attempt to imagine a utopia where the governing structure has withered away and everyone lives in blissful equality with no single individual or group seeking to represent the whole; but from the point of view of survival against a deadly environment, it is unlikely that such a state of affairs is optimal. Even with a well-organised egalitarian participatory democracy, it is desirable for certain individuals to be endowed with the responsibility to act on behalf of the settlement in emergency decision-making; this is, in an early state, a type of executive branch.

Furthermore, a clearly defined governance structure with organisational clarity is a benefit to liberty; authority becomes abstract and obscure in diffuse power systems without clear centres.[50] The capacity for individuals to enjoy a realm of non-interference is not only brought about by a lack of coercion from other individuals; this duress can come from a nebulous structure of a state or set of governing institutions. Clarity of the governance structure is therefore essential if people in an extraterrestrial setting are to identify sources of coercion and seek to rectify them. A lucidity in the organisational hierarchy of governance structure is the first step to identifying the inevitable tendency that organisations have of expanding the remit of their action.[51] We could therefore observe that whatever government is constructed in space, its terms must be plainly stated (for example, in a constitution) and its boundaries and modes of operation and responsibilities intelligibly laid out. The question becomes how this power is to be managed, not whether we can dispense with it entirely.

Another attribute that we might observe, which derives from the considerable pressure that exists for a form of highly centralised organisation against the external environmental conditions, is that any governance structure we imagine should conform to the general ethos that it only has responsibilities to the people it serves.[52] Once it sees itself as being the bearer of rights to be demanded by a people corralled under it, and considers this function its operating principle, its privilege and reason for existence, it is already an instrument of tyranny and coercion.[53]

Within the inescapable need for collective coordination, we could therefore say that the general end state we wish to drive towards is an organisational structure that views itself as a servant to the people. It governs in such a way as to allow individuals a private sphere of non-interference and the capacity to express their views (a freedom of expression) and believe the things they want to believe (a freedom of conscience). This is not a bland statement of liberalism, but these characteristics have an immediate practical benefit in generating a people with independence of mind from whom creativity and solutions to the settlement's challenges are more likely to be forthcoming. One does not need an ideological affiliation with liberal views to see that individuals who have the capacity to disagree, dissent, and think are likely to be better equipped to survive, and to aid survival of their fellow human beings, in an environment where many needs of everyday existence must be extracted from deadly surroundings. The reader might counter this statement with the opposite argument, that lethal conditions rather encourage a highly hierarchical structure that discourages unpredictable dissent. However, I am referring here to a long-term society in which people spend considerable portions of their time, not the sorts of short-term societies found, for example, in Antarctic stations. If we wish for a society in which innovation and originality can have a field of play over long periods, and an inflexible order is not dominant, then I contend that an extraterrestrial liberal order is preferable. The extraterrestrial governance structure must therefore protect a constellation of values which are as much use to the individual as they will be in creating the space for individuals to realise the ideas and fully developed personalities necessary for the collective success of a settlement.[54]

With these ideas in mind, I will propose that forms of governance that involve the population in decision-making to the maximum extent possible, in other words having a democratic element to their constitution, are highly supportive of these aforementioned objectives.[55] It is not that a centrally planned undemocratic system cannot recognise freedom of speech or conscience, or even see itself as a servant to the people, it is that human experience has shown that these values are generally less likely to be recognised under such a system.[56] When we plant a system of government in a location where one of its chief objectives is to protect against instant death, we can see that all of the problems I have brought to attention in the previous chapter are more likely to be realised when the governing powers have the security of never being subject to popular judgement and dismissal. This, then, is the essence of the political liberty to which I will direct attention through this work. As Arendt averred, 'For political freedom, generally speaking, means the right "to be a participator in government," or it means nothing'.[57]

Furthermore, I would not deny the possibility of a benevolent paternalistic arrangement where the settlers are cared for by an authority that shows an immense and unfaltering sense of responsibility for their welfare in extremes and where the settlers themselves submit to this design. If the relationship between the settlers and their management structures can be established as congenial from the start, then this guardian-like autocracy has every chance of success. However, these structures are likely to fail to energise the settlers into an active interest and participation in their polity, leading to a lethargic dependency that is, or becomes, to all intents and purposes, a master–servant relationship. Such a relationship, although viable in that it might satisfactorily survive, is unlikely to be conducive to the immense creative capacities of the human mind that will enhance the success of space settlement, or to releasing the full scientific, cultural, political, and economic potential of space societies.

These, then, constitute just some of the essential characteristics of a free settlement: an authority that can be effective, yet subject to popular censure and recall, and within which individuals can find a space for their personal liberty. What structures of government or organisation might best realise these goals as a settlement develops?

Participatory democracy in the early stages of settlement

The nature of political power in space will be linked to the size of the population. Most extraterrestrial settlements will begin small—maybe a few tens of individuals—but there is potential for some of these to expand beyond localised science or tourist stations to larger settlements that might exceed thousands of people or more. What sort of government structures might recommend themselves to these settlements in the early stages?

For a small settlement, a direct form of participatory democracy seems feasible.[58] On Earth, there are countless examples of small societies run by participatory democracies, and a long history of diverse types of collective decision-making is available for consideration. Even virtual worlds created on the internet can yield insights into human social organisation with potential application to space.[59]

A modern example might be useful to illustrate the possibility. In the Scottish Hebrides are a range of island populations that have experimented with diverse governance structures. A case in point is the Scottish Isle of Eigg. The island lies about 16 kilometres off the west coast of Scotland and is approximately 8 kilometres by 5 kilometres in size. It has hosted human populations probably since the end of the Neolithic. Over a few years of study, I have given this island some attention with respect to understanding how its governance structures can inform the political philosophy of space settlement.

Fed up with decades of absentee lairds (the rulers of Scottish islands) and their unpredictability, the community took ownership of their island in 1997 by purchasing it through an organisation they set up: the Isle of Eigg Heritage Trust. They then established a participatory democracy. The island has a population of approximately a hundred people, which makes participatory democracy practical. The community hall provides a venue for meetings to discuss matters of general importance to the whole community.

Overall planning of the community is undertaken by the Trust, which has three members: the Isle of Eigg Residents' Association, the Highland Council, and the Scottish Wildlife Trust. Mapped into the extraterrestrial environment, the Residents' Association would correspond to all the people living in the settlement and would be its participatory democracy. The extraterrestrial analogues for the Highland Council and the Scottish Wildlife Trust would be any form of organisation on Earth that can provide additional care and influence on behalf of the isolated settlement, for example economic or trading organisations that provide useful commodities from Earth or represent the settlement on Earth. The mode of their involvement I will discuss shortly. The representation of external organisations would be weakened or non-existent in a settlement that had achieved complete self-sufficiency and independence, yet even the most independent settlement may still wish for some involvement and collaboration from organisations outside the settlement, such as Earth, for economic and political purposes.

Participatory democracies, despite their egalitarian image, do have weaknesses,[60] and one should not be so naïve as to claim that any process that is democratic is by default good, without any further disputation.[61] They are subject to the fashions of the crowd or to influential demagogues, one of the perpetual flaws in the democracy of ancient Athens.[62] 'In the courts of princes, men whisper their flattery; in the midst of a democracy, they flatter the people to the sound of a trumpet' as Necker memorably suggested.[63] We might also agree with Burke[64] that liberty can be a passionate thing and sometimes needs mollification, but this need not tacitly imply, as he was inclined to, that the power to reduce these passions need be wielded by an unelected force outside of the democratic institution.

The caprice of tyrants, over-zealous passions, and demagoguery have not been witnessed on Eigg in any detrimental way (although Eigg does not assemble armies and armadas, so the effects of these frailties in that democracy are less likely to be recorded and are of less historical consequence), although there are individuals on the island who become dominant in matters, but mainly because of their enthusiasm and energy and not because of nefarious intent.

A mechanism to dilute the unpredictable tendencies of a direct democracy is to create a form of mixed constitution, whereby the democratic assembly is balanced by an executive council that may still be elected but considers matters in a more detached mode from the assembly as a whole. The idea that governmental structures are most stable when they incorporate mixed forms of deliberation to dull both the excesses of mob rule in democracy and the dangers of the usurpation of power in aristocratic or monarchic modes is an old insight, expounded by Aristotle, popularised by Polybius and Cicero,[65] and later eloquently discussed by Bolingbroke.[66] Indeed, inspired by the classical deliberations, the founding fathers of the United States went to great pains to avoid a democracy but instead to conceive of a republican system with a mixed constitutional design. If we strive for stability in the extraterrestrial case, it would be wise to heed the lessons of history and to attempt to devise forms of governance that seek the balancing influence of power centres, even in the earliest stages. Although a simple participatory democracy may be attractive, since it minimises bureaucracy and political embellishments, additional complexity added by executive functions may act to improve the chances of long-term success, not cause destabilising complications. We wish to avoid what Polybius observed about the participatory democracy of ancient Athens:[67]

> the best analogy for the Athenian democracy is a ship without a captain. On such a ship, the crew do their duty outstandingly well as long as fear of the open sea or the threat of a storm induces them to cooperate with one another and obey

> the helmsman. But when there is no cause for alarm, they start to ignore their superiors and to fall out with one another. And then the infighting begins: some of them want to continue the voyage, while others urge the helmsman to drop anchor; some of them let out the sheets, while others interfere and order the sails furled. It is not just that this quarrelling and fighting disgust people looking on from outside; they also make things dangerous for everyone else on board the ship for the duration of the voyage. And what happens is that often, even after crossing the widest stretches of open sea and surviving the most furious storms, they founder in harbour or close to land. This is certainly what has happened on several occasions to the Athenian state.

Returning to the Isle of Eigg, the island's Trust is overseen by a board made up of four directors who are elected by the islanders and two directors provided by the Highland Council and the Scottish Wildlife Trust (one director each). The board meets quarterly, while daily management is entrusted to the island directors, who meet on a monthly basis. In a simplistic way, the island directors might be viewed as an executive branch of their democracy, which provides a steer and structure to deliberations of the community at large, a captain for the ship, and may be a significant contributor to the stability and success of that democracy.

Transplanted into space, gatherings of the whole community could be presided over by elected officers who are re-elected on a regular basis; in this way, a nascent executive branch consisting of a president or chair with a deputy and two to three assistants would represent this management function. Like Eigg's relation to the Scottish mainland, this group could be augmented by representatives from Earth or other external organisations. We note that in the case of Eigg, the island, with its four elected directors, constitutes the majority in this executive function, with two directors from the mainland. Similarly, in space, any executive body should be majority controlled by the settlers themselves, as they are the ones who must face, and respond to, the periodic challenges in their environment.

Although I will return to economic activity, in the context of the wider Eigg society it is of value to point out how this part of their community functions in terms of the analogy I am drawing. As might also be replicated in space, the Isle of Eigg Heritage Trust has subsidiary companies, including Eigg Construction to oversee major building works, Eigg Trading to handle the store and amenities that cater to tourists, other visitors, and permanent residents, and Eigg Electric to manage the island's electrical system. We could imagine similar arrangements to organise the economic affairs of an extraterrestrial settlement, although, as I will elaborate, I will suggest that monopolies in economic activity should be avoided, with plurality conducive to the decentralisation of control, particularly in vital resources. Mirrored on the Moon, for instance, one might imagine Moon Construction to oversee construction and infrastructure works. Moon Trading might be a collective of organisations involved in facilitating all forms of economic trading both on the Moon and beyond it. Moon Electric might oversee the power system. We might also imagine similar associative organisations, comprised themselves of a variety of independently operating organisations, dealing with oxygen, food, and water.

These economic activities underline another principle that the decisions of a society should not all be run through a single participatory democratic process and its executive branch. Voluntary association in economic activities and other arenas provides multiple avenues for the expression of ideas and human capacities. In the case of Eigg, we note the separate trading organisations that work with the Trust, but offer a place for the expression of alternative economic views. In the lexicon of Elinor Ostrom, this 'polycentric'[68] organisation makes at once for diverse, overlapping centres of organisation, diluting the tendency for intense centralisation. These organisations have a propensity to soften the caprice, passions, and potential mob rule of a society where all decisions emanate from a single participatory democratic congregation.

Multiple centres of organisation dampen yet another weakness of democracy. In contrast to the tendency to unleash uncontrolled passions, democracy at all scales can have the countervailing effect of causing conformity, the very problem we have identified as a special concern in the extraterrestrial environment. Demanding of a general culture of equality of treatment, one of the laudable aspirations of the democratic mindset, this very outlook can demand the unyielding implementation of majority view to the detriment of the minority.[69] The tyranny of the majority then becomes the antithesis of the unpredictable democratic mob, instead encouraging the predictable conformity of the crowd bearing down on dissent. In the previous chapter, I gave empirical evidence for this latter problem in the Scottish islands. The voluntary associations of Eigg provide a template for dispersing this overbearing conformity and allowing conduits for individuality through associative activities.

Finally, a prominently perceived weakness of democratic procedures is the time taken in wrangling, discussion, and debate, which can make the participatory democracy, indeed any democratic form of government, less efficient[70]—a problem that could become acute when dealing with distinctly extraterrestrial crises such as a depressurisation or a fire. However, this is a pale attack on democracy for the reason that this inefficiency is precisely what democracy is about: removing arbitrary or inflexible rule and replacing it with the messy, uncertain, but consensual mechanisms of involving the people in the governing of their society. Furthermore, the presence of this deliberative slothfulness does not negate an executive branch taking rapid decisions under crises in extraterrestrial conditions, decisions that are later subject to democratic scrutiny. This mechanism is well known in democratic societies on Earth when faced with exigent emergencies. On Eigg, the directors can address emergent crises of this type in an effective and focused way that would not be considered by anyone to be despotic.

Thus, we may conclude that democratic government can be realistically created in the extraterrestrial case, and in the smallest communities; but a simplistic resort to a pure participatory democracy would likely be a great mistake. Lessons extracted from small permanent communities on Earth can be used to fashion a modified participatory democracy in which the known frailties of this system of governance can be minimised by the use of mixed constitutional designs and voluntary associative capacities.

Unelected positions in the highest positions of authority should be avoided

Leading from the general question of the democratic means of social deliberation is the matter of whether one might include unelected positions at the highest levels of the governance structure, even if that structure is on the whole democratic. Whatever system of governance is chosen, I believe it to be of benefit to the society that the individuals at the top of the edifice should be subject to popular censure and election to ensure that they do not usurp the power of the settlement. It is not that an unelected leader cannot be benevolent, but that, consistent with Madison's observation that people are not angels,[71] the population is forever in debt to those individuals for not overreaching their powers, even if in theory they are restrained by rules. When a people must depend on the benevolence of their leaders not to be subject to arbitrary rule, they are as good as slaves.[72] For this reason, a system of government where the leaders are unelected, such as monarchies, but in fact any unelected structures, should be discouraged.[73]

I do not doubt that through the vicissitudes of history it is possible to create a system where the highest position in a governance structure is unelected and whose power is sufficiently shackled such that it is not despotic, as is observed in the case of the British monarchical system of the state.[74] But here we are relying on a propitious set of historical and political circumstances that happen to keep an unelected head of state properly controlled with respect to the power that they wield.[75] To set out to install an unelected head of state and to hope that artificial mechanisms created to limit their power will be effective in the conditions to be found is surely to flirt with autocracy, particularly since if the mechanisms of control are ineffective, it will be difficult to remove an unelected figure.

The observant will point out that unelected leaders are to be found in a variety of communities living under extreme conditions. Antarctic stations do not elect their commanders, and nor do space stations; they are selected by the organisations running the outpost. However, the crucial point of distinction here is that these are not permanent settlements, and the leader takes command only temporarily—for a season in Antarctica or a single mission on a space station. Dictatorial tendencies, if they emerge, will therefore hold sway only for short periods. However, what we are considering here is an attempt to emplace a permanent population with governance structures that can yield to self-sufficient societies in locations a long way from Earth. Even if we were to instigate a regime where the commander of a settlement was not elected but was removed after given periods of service, we will have lost the opportunity to generate a participatory and responsive political culture. The election of the highest posts in the settlement is a crucial step in building functioning democratic institutions that are accountable to the settlers and in which the settlers themselves feel a strong measure of involvement and direction. In the long term, a society of confident and free-thinking citizens is more likely to emerge from such conditions.

The extraterrestrial society, at its inception, will be revolutionary

It might be a point that is obvious to the reader, but it is worth drawing attention to the fact that the extraterrestrial society, particularly upon its immediate inception, will be revolutionary in character. It will be revolutionary not because its founders are people of revolutionary aspirations (although that may well be the case), but because it will be planted into an extraterrestrial environment where no previous society has existed before. It will be inescapably revolutionary from its very beginnings.

The relevance of this quite uncontroversial observation is that many of the arguments and counterclaims that we have seen on the field of liberty will be initially irrelevant to this society. Edmund Burke's defence of the slow evolutionary development of British society based on customs and traditions, as opposed to the recent revolutionary changes he observed in France, cannot apply in space, since there is no prior extraterrestrial society from which to slowly evolve, or from whose institutions improved organisations can emerge. There is no superstructure of customs, traditions, mores, and rules that exist unformalised in the body politic from previous ages. Chesterton's fence cannot be an object of concern in an unpeopled barren landscape.[76]

Of course, it is true that this society will take some precedent from terrestrial laws and rules; it will not be wholly revolutionary in a pure sense. Even the most original extraterrestrial societies will have embodied within them knowledge and experience projected from Earth. But then one might make this observation of even the most revolutionary societies on Earth. The Jacobin upheaval in France contained strands of the *ancient régime* within it, however unrecognisable they may have seemed. When we talk about a society being revolutionary, what we mean is that the society undergoes a tectonic rupture in many of its most fundamental institutions and traditions of government, that there is a sense that the old regime is being swept away and replaced by a new type of society. This is what we generally mean when we speak of revolutionary societies, although none of us can place some quantitative statement on the number of institutions that must be modified for us to declare that a society is undergoing revolutionary as opposed to evolutionary change.[77]

Perhaps this extraterrestrial revolutionary ardour will be most forcefully felt in the definition of rules by any governing power. Although they will extract precedent for rules (which may even be called 'laws' or 'rights' at this early stage) and their implementation from terrestrial analogues, there will be a range of specifically extraterrestrial activities that relate to oxygen, spacesuits, habitats, and other paraphernalia that will bring forward the need for regulations unseen on Earth. This will gradually change once societies do emerge and they can become a template for later iterations either in the location in which they exist, or for societies being established in similar environmental conditions elsewhere on the same planetary body. However, even if similar legislative instruments have been established already in space, each new location will require a particular manifestation and interpretation of those rules.

Seen in legal terms, this novelty in the space frontier will drive society strongly towards the development of a codified form of rules (similar to a civil law), since customary practices that may provide the basis of something akin to a common law will not yet exist, or at most they will exist in rudimentary forms with few prior examples. In this condition, society will be tilted towards a revolutionary state of mind in that an entire series of new rules and their interpretation must spring immediately into action to serve a new branch of humanity.[78]

This aspect of revolutionary character will be given succour by the need to do other things that we associate with revolutionary societies, such as defining new calendars and markings of time, made necessary by the orbital mechanics of places distant from Earth.

The point germane to this discussion is that revolutionary societies tend to risk discord and social disintegration, violence even, when entirely new institutions do not mesh with each other. In old societies, the centuries of prior cohabitation of institutions that leads them to settle down into a jigsaw-like arrangement where they modify one another as social conditions change is disrupted by revolution. Before a new system and web of governance emerges from the chaos, there is a chance for social convulsion of a very destructive kind. In the extraterrestrial case, the society will enter into the stage of history as if it had appeared at the beginning of a new revolutionary era. The institutions must settle into a mutually supportive framework almost immediately. While this process occurs, the frictions between these novel institutions have the potential to lead to disruptions, coercions, intense argumentation, and even violence.

It will be necessary for the new extraterrestrial population to accept that this historical stage cannot be avoided. They will be constructing a revolutionary society, and the misalignment of new institutions and the discomfort that they bring about is an inevitability. What is therefore called for is a forbearance and a lack of revolutionary dogmatism among the founders of the new society. They must work towards the achievement of a democratically decided optimisation of institutions, making deliberate and careful analysis and consideration of the institutional arrangements and how they might be improved as the social order solidifies and settles. This new form will never be perfect by any single individual's estimation. The potential for destabilisation occurs when an individual, or group of individuals, has a fixed and unyielding view of what that first flawless form should be and any deviation from it, merits, in their mind, new social turmoil.

The forbearance that I draw attention to here is necessary in view of the fact that not only will the society be revolutionary, but its early developmental phases might occur in rapid succession according to an unfolding history that has never before been witnessed. What we have before us is a society that starts from scratch in a revolutionary phase, creates a system that might be something akin to an ancient Athenian participatory democracy, and then, as it grows, transitions to some type of republican system of representative government. Whatever the exact flavour of these transitions and intermediate stages, there is the potential here for, at least in conventional historical terms, a rapid numerical expansion in people and concomitant social metamorphosis which, when combined with the necessary revolutionary element, could lead to immense instability.

Need it be stated that the extraterrestrial society should be vigilant against this inescapable revolutionary phase becoming an excuse for violence? The Jacobins, and before them Machiavelli, were utterly comfortable with the idea that the founding of new republics involved necessary violence. As Hannah Arendt observed, summarising their views: 'You cannot make a table without killing trees, you cannot make an omelette without breaking eggs, you cannot make a republic without killing people.'[79] Yet violence is not a necessary ingredient in the recipe of a new political and economic order; there is no natural law that says that people must die to make republics. A better analogy with trees and omelettes would be to observe that one must surely break down old institutions and rules to make new ones, but not people. Regardless, in the extraterrestrial case, there is no old order to disband; one is erecting new structures from scratch, one has been handed the wood and the peeled eggs. Institutional destructiveness, let alone the annihilation of people, is wholly avoidable. From a purely pragmatic perspective, in the extraterrestrial case the inception of violence may be terminal for a new republic. Violence in a young and barely sustainable outpost where the infrastructure is still inchoate will be not the birth pang of a new order, but a signal of a stillborn society.

I cannot say what imperfect form should be considered acceptable by the population as a starting point for the society, although in this work I seek to set out the general ethos of the totality of that society. I think it is sufficient to say that unless the new institutions settle down into an undeniably autocratic state where there is a widespread sense of coercion and arbitrary domination, the extremes of the extraterrestrial environment would suggest that continuous cycles of social tumult should be avoided; a slightly ramshackle and unsightly democracy is better than further rounds of instability and the risk of despotisms, particularly when there is no clear certainty that the new arrangements will be significantly better.

On the transition to more representative modes of governance as a settlement grows, and observations on the means of pre-emptive free governance

As a settlement's population expands, it is the case that participatory democracy becomes more impractical to implement and deliberation must move to a more representative mode.[80] It is not possible to state an exact number at which such a transition occurs. A participatory democracy of several thousand was possible in the open spaces of ancient Athens, but its weaknesses and tendency to be dominated by battles between particularly loquacious individuals is well known. The optimal size at which a transition from participatory to representative democracy should occur is a matter of considerable theoretical and practical interest and worthy of further analysis. If we consider the pragmatic matter of assembling everyone in one place in the confined spaces of an extraterrestrial settlement, combined with consideration of a number beyond which individual voices tend to be drowned, this might suggest that between approximately two hundred and a thousand people, a society is probably ripe for a representative system. One tends to see this type of transition in governance in Scottish islands that reach these numbers. There is no critical threshold at which a sudden and marked change of governance occurs; as populations increase, fewer of the decisions involve the whole island. Even on the Isle of Eigg, with a population of barely more than 100, many everyday decisions tend to be implemented by elected directors in a form of representative executive branch.

These transitions, however, might be accompanied by formal changes in the way of an explicit resolution to move from decision-making entailing the whole community to electing individuals to represent the interests of the settlement. At even later stages of settlement growth, let us say settlements of a thousand individuals or more, areas of the settlement, which might have developed their own character, might elect representatives in a geographically representative fashion.

I do not think it is possible, or even desirable, to give scientific force to the matter of the numerical thresholds at which these political transitions should occur; they will depend upon the physical layout of the settlement, its political culture, the political involvement and energy of its population, and other factors. However, from its inception, the settlement might actively keep the intention of completing these alterations within its view. When the participatory democracy starts to creak and shows the weaknesses and growth pangs that make it inefficient, it will likely become naturally clear to the population that the time has come for more representative arrangements to reduce the stress on the system of governance.

It has often been thought that the growth of a settlement to a size that no longer admits of participatory democracy will necessarily lead to an erosion of liberty on account of the dilution of individual voices by a larger franchise.[81] However, this need not necessarily be the case if the structures of a governing edifice contain checks on power and if the society effectively encourages the expression of diverse individual views outside these structures, such as in largely independent voluntary associations.

There can be little hesitation in observing, though, that a transition in scale from a small settlement to a much larger one requires a fine balance between the power of the individuals in society and the power of the authorities overseeing that settlement. Too much influence of the former may lead to a disintegration of social order, which would likely manifest itself if some catastrophic depressurisation or lack of provision of oxygen, water, food, or power occurred from a lack of coordination; too much influence of the latter will lead to rigid, unsympathetic power structures. In any society, as it changes or grows, this balancing act is extraordinarily difficult to maintain. It has been examined for the terrestrial case in erudite detail.[82] In reality, we do not know, and cannot predict with scientific confidence, what factors allow for the success of democracy. If we were to accept the recent historical conclusions of Stasavage[83] that democracies have tended to emerge in areas of the world where geographical and other factors have forced upon initially weak states a requirement to seek the consent of the people, for example in areas where individuals are widely dispersed and have little need to obey a weak state bureaucracy, then we might be concerned for the extraterrestrial case, where people will be highly localised and living within an administrative structure that for reasons of survival will have to be strong from the outset.

Fenced into habitats from which there is no easy escape, those who command the settlement have little need to ask the people with an air of obeisance for their right to govern. In Stasavage's analysis, the extraterrestrial situation, comprised as it will be of a small population whose members are utterly dependent on each other and the organisation of the technology that sustains them, should be ripe for autocracy. This type of conclusion, whether we believe it or not, lends support to the idea that in the absence of tools to predict the conditions, and required actions, to create democratic institutions, we should take the Madisonian path of assuming that humans will *likely* fall prey to despotism, mandating that we think of all the ways we can to prevent it. In the extraterrestrial case, we might emphasise that the people must be vigilant to their institutions as these transitions occur and never grow complacent by assuming that an effective participatory democracy in an early settlement will naturally transition to an equally free larger representative structure. What may begin as good intentions in improving oxygen production can lead to totalism through goodness, not maleficence.[84] Indeed, my purpose in this work is to suggest some of these mechanisms.

As a settlement transitions into a more comprehensive representative mode, what was the participatory democracy involving the whole settlement essentially becomes the legislature which is established by election. Those individuals who played a role in implementing the decisions of the settlement become defined formally as the executive. In a primitive way, the branches of the government, through the separation of powers, becomes a republican government; it remains democratic in its methods, but essentially republican in its electoral mechanisms and the systems that represent its power.[85]

In the extraterrestrial case, the transition from small participatory democracy to larger settlement could be relatively quick, at least in comparison to many settlements on Earth, and a smooth transition between these forms of governance appropriate to a few people and many will have to be achieved. One potential mechanism I have suggested as a means to attempt to instantiate the condition for liberty within political structures as they grow is the idea of 'pre-emptive free governance'.[86]

To prevent a disordered transition from a participatory democracy to a more representative style, the settlers might pre-emptively produce a separation of powers from an early stage. In a small settlement, we might have a group of three to four people charged with overseeing the running of the settlement (they are essentially the executive, as we see on Eigg, for instance), but there will be separate representatives (again, three to four people) who deliberate the laws and rules by which the settlement will be run (the legislature). To begin with, these individuals are small groups who are part of the participatory democracy that constitutes the whole settlement, but they are elected representatives of the whole settlement. Then there is a small body of people who are responsible for implementing the laws whose boundaries may be circumscribed by a constitution; they are the judiciary. In a small settlement, this juridical function may be a single individual who is a police officer.

Consider an example of this modus operandi of pre-emptive free governance and the culture that it seeks to establish—the formulation of rules (nascent 'laws') in an early settlement. The job of the primitive legislative 'branch' of governance would be to develop the cultures of the rule of law and equality before the law, wherein the rules established across the settlement are applicable to all. In a small space community, we could begin to achieve this by focusing on deliberately creating rules (for we can hardly call them 'laws' in these relatively small, informal circumstances) that are applied equally to all people in the settlement, establishing the cultural legacy leading to a rule of law and equality before the law in later stages. For example, I have been in Antarctic stations where rules are established but there is no deliberate attempt to create a culture of the rule of law; there might be a rule for scientists, or a rule for support staff. One such rule could be: 'All scientists on base must clean up their laboratory waste'. Although we can accept that there must be procedures specific to particular people (a rule about cleaning up laboratory waste has no relevance to a chef), we could purposefully attempt as far as possible to generate rules that all people are subject to without specificity—in order words, the progenitors of laws. 'All waste in individual rooms will be cleaned up' now applies to scientists and chefs: it targets no specific group; it is akin to a law under which all people are treated equally. More than this, the deliberate attempt to generate general rules that amount to laws could be an overt objective, discussed among the settlers in meetings convened for this purpose and informally during meal times, thus inculcating a culture of law development that all new settlers would be involved in and subject to. Attempting to generalise rules in this way is a sociable activity, and could be a source of enjoyment for early settlers seeking ways to play a full and involved role in the development of their polity. If it involves everyone, it may have a persuasive, enriching, and lasting effect on the mitigation of tyrannical tendencies and individuals, and a positive effect on social cohesion and deliberation.

Implemented immediately in a small settlement of tens of people, this suggested structure might seem comical, since much overlap between the 'branches' of government would occur and their numerical scale would make them seem unnecessarily formalised. Yet the purpose of such an arrangement would be to incorporate the concepts of the separation of powers into governance, and the checks and balances that can be explored between them, at the earliest stages and ensure a more fluid and ordered transition to a larger form of representative government at later stages. Even in the smallest participatory democracy, this early separation of powers may aid in avoiding the well-known weaknesses of the tyranny of the majority or mob rule, while bringing the benefits of the separation of powers found in larger representative democracy.[87] I have shown some evidence of this on the Isle of Eigg.

We should not convince ourselves that the separation of powers, in itself, can guarantee liberties: it can, if it fragments government too greatly, create weak and anarchic government. In extremity, that is as much a threat to the effective operation of a government, and an invitation to tyranny, as it is a check on power.[88] Indeed, the separation of powers is an unfortunate, although now hackneyed and established, phrase, since few political philosophers have ever suggested that the branches of government should be categorically separated. Jacques Necker, in particular, was most fervent about the need for the branches to be deeply intertwined in such a way that they maintain a vigorous power and independence, but at the same time argued that the caprice and authority of each must be kept in check.[89] Yet the general ethos that these suggestions entail—that the settlement must attempt to avoid political arrangements whereby the population is corralled under the control of a centralised group of people over which they have little influence at any point along its developmental trajectory and through which their voices cannot be heard—is a serious one.

Pre-emptive attempts to discourage autocratic modes of development as a settlement enlarges are given special force by the likelihood that early settlements will tend to have somewhat centralising structures, since production capacity will be low and effective organisation and hierarchy necessary. This would be the case, by default, in any settlement established with a strict mission and focus, but the most deliberately laissez-faire type of arrangement will, even with a participatory democracy, begin with a centralising tendency.[90] The mechanism of pre-emptive free governance is just one suggestion for attempting to create democratic structures from the outset that can develop and complexify, acting as a counterweight to this centralising habit.

A quality within the notion of pre-emptive free governance is to encourage a stable evolutionary transition in governance structures as a society progresses. I have noted already that the society will necessarily begin as revolutionary, but even after its establishment, this essential character is bound to linger. A gradual growth and development seeks to avoid revolutionary ruptures in the governance of an extraterrestrial settlement as it develops, with all the dangers that entails. In the lethal confines of space, although the society starts off as revolutionary, we then seek a rational evolutionary social development that is distinctly Burkeian[91] in its flavour. Edmund Burke was a strong proponent of monarchy, but here I mean instead to draw attention to his view of the necessity of pragmatic institutions that take into account history and tradition and evolve in that way.

The regular disruption of extraterrestrial governance structures by revolutionary overhaul and redesign is tempting for those who might perceive this to be more 'scientific', but it is to invite disaster in the extraterrestrial environment. It is likely a general point of fact that if you tear apart the existing social fabric and the political and economic structure of a society with the intention of completely reshaping it, you are probably more likely to cause social disorder, disintegration even, than if you attempt to change structures around people piecemeal.

Given the relative tabula rasa that extraterrestrial environments will offer to new forms of social organisation, there is a strong temptation to pursue rational designs of society at all stages of development, even after that more obvious initial revolutionary phase, much like an architect might redesign a building and on occasion recommend demolition and redesign. To pursue this architectural analogy, what we are seeking to do is to design a building that can be restructured, added to, and modified, yet whose essential foundations and walls can sustain changes over time without the need for total demolition. This is the essential political challenge for settlers beyond Earth, and it suggests that early structures should be designed with later modification in mind. This is the substance of the suggestion of pre-emptive free governance.

Some observations on the branches of a maturing extraterrestrial government

A discussion on the branches of government may seem a little ambitious and far-fetched in view of our expectation that stations on other planetary bodies will, at least initially, come nowhere near the scale of national states on Earth. Yet consideration of these potential branches and the manner in which we might want them to operate can inform the earliest societies about the general political trajectory on which they might seek to travel. For these reasons, it is of value to make some general observations on how, if we were to operate a large enough settlement where a separation of powers would be used, these branches might be run and how they would relate to each other.

I will assume that the branches of governance that are of interest to us are executive, legislative, and judiciary.[92] This is not because I slavishly follow modern terrestrial precedent: maybe there are entirely new forms of government divisions that I lack the imagination to conceive.[93] However, I think that these forms have a certain logic to them. Considered from an ahistorical perspective, we could say that we need a branch to run a society in its ceaseless activities, a branch to set rules, and a branch to interpret and enforce the rules. Beyond these essential functions, all else can be done by subdivision. The existence of executive, legislative, and judiciary branches in terrestrial government therefore seems not so much a product of historical political contingency, but rather a reasonable reflection of universal requirements of human management at the large scale whose discovery has emerged from millennia of human social organisation and experiment.

As the rules under which people live can be an acute point of disagreement in any society, a singular point of importance in devising a successful means to expand a settlement is to consider how the rules are to be determined as a settlement becomes less directly participatory. In essence, we are discussing the means by which a legislature, which could in the early stages of representative government be a small body of people, can modify itself as a population grows.

There are a variety of modes that we can imagine by which the legislative functions of a settlement could be implemented in this later representative stage. Settlement rules could be debated and discussed by a unicameral body, but those rules could be distributed to the whole settlement by electronic communications for their vote. The process might yield an unwieldy number of these plebiscites, in which case certain categories of rules, such as those pertaining to oxygen, water, and food production, which can have far-reaching implications for people, might be prioritised as being subject to popular democratic censure with other decisions taken and implemented by the legislature without further recourse to the populace. In this event, the legislature would transition to a more local representative mode of decision-making, yet its decisions could still be subject to veto by the population, for example by the mechanism of the collection of a certain number of votes to challenge legislative decisions.

An alternative structure that has proven itself in a number of countries on Earth is a bicameral legislature. A bicameral structure can complicate considerably political decisions, but one of its major advantages is that it tempers the tendency for a unicameral body, in being the sole disperser of rules, to acquire an aloof and unchallenged imperiousness in that task.[94] We could go further to dilute power: a tricameral structure would probably dilute this tendency by an even greater extent, yet it would likely generate an unnecessarily convoluted process (it is difficult to imagine what that process would be). Many terrestrial societies have settled on a bicameral structure as a viable compromise. A bicameral arrangement can allow for checks between both houses without burdensome intricacy. Some observations about how such a structure might work beyond Earth are worthwhile.

One house of this bicameral arrangement could be formed of elected officials of the settlement as a whole, who are charged with representing the local and immediate interests of the populace. The second house might be composed of a collection of individuals, also elected by the settlement, who have extensive and deep knowledge of the various political and economic issues facing the settlement at the largest scale. We might think of this as a house of the wise.[95] In one mode of operation, the house of representatives debates rules and laws that are then considered and voted upon by the second house, who may reject or demand amendments. This system of operation bears some similarities to the British parliamentary system.

An alternative mode is that the second house debates rules, regulations, and laws, and the results of those deliberations are then further debated on and voted by the first house of representatives. In an environment where lethal conditions might overwhelm a settlement, there is much to be said of the second approach, whereby the house of wise, dispassionate representatives considers the rules required to safely operate the society, but these decisions are censured by the first house of representatives, who can determine if those now a priori wise rules comport with the general mood and political heartbeat of the settlement. Such a process may dull the propensity for immediate passions to endanger the settlement. However, to ensure that the people of the settlement do not live under the tutelage of the second house of the wise, the first house of representatives should have the capacity to bring forward proposed new rules and laws that address immediate concerns. A mixed form of representative and direct democracy, similar to the Swiss model, could be incorporated, wherein the people themselves, after gathering a certain number of signatures from their fellow citizens, can force upon the legislative the consideration of new laws or the repeal of existing ones. Thus, the population as a whole exerts a direct democratic direction and veto upon the legislative activities of the settlement and their representatives.

A point worth drawing attention to here is that I have not suggested a legislative structure similar to that in the US or Swiss models, in which one house of a bicameral structure is comprised of a fixed number of representatives for a given geographical area and the other a number of representatives proportional to the population of a given area. This is another potentially effective model, but it requires a society in which the size of the population and the variety of geographical segments of the physical settlement make it a plausible organisational model. Eventually, if we could imagine an extraterrestrial society—for example, on Mars— constructed from districts with their own local concerns and political and economic problems, much like states in the US or cantons in Switzerland, then perhaps a legislature could transition to such a form. Prior to this scale of settlement, however, a legislature that allows for a bicameral check and balance, yet does not have the possibility of a separable populace across large geographical areas, might suggest the greater practicability of non-geographically based legislative houses.

Speculating on the fine-grained structure of a legislature for a new society is entertaining, but also potentially open-ended in an unproductive way. However, in whatever form we imagine such a legislature, I think we can see several points of importance that are specific to the extraterrestrial case. First, it is essential that there is a house of representatives that reflects the needs and wants of the people at all times and mirrors their changing moods and thoughts. This would suggest regular democratic elections to ensure the correspondence between the people and their representatives; this is made essential by the danger of the extraterrestrial condition and the need for the legislature to be rapidly responsive to it. These representatives should embody the will of the people who elected them, but in the extremities of the extraterrestrial environment, when gathered to deliberate the conditions of the settlement, they should possess the independence of mind to come to agreement with their fellow representatives on a good course of action.

Second, we can see the value of a second house of cool heads who know about the oxygen, water, food, power, and other productive capacities in a settlement and can, in a calm, safe, and deliberate manner, free of wanton passions, discuss and debate proposed new rules. Facing potentially dangerous crises, this branch of the governing powers, which acts as a brake on capricious decisions, may be of great benefit.

Third, we can see that interactions between these two groups, in a way whereby neither dominates the other, is essential, and this speaks to us of the value of a bicameral structure with checks and balances between the two groups, each aiding one another. Both houses operating together combine the vivacity, energy, and liveliness of the common opinion with the considered, knowledgeable, and calm deliberation of that portion of the body politic that embodies these characteristics.

I believe the number of individuals elected to different houses is not something that can be mathematically formulated prior to the formation of such settlements; it will depend on the size of settlement, the nature and forms of expertise in the population and its character, and the dangers and challenges faced by any given settlement. Nevertheless, we can apprehend that, these details aside, the general conditions for liberty in extremity demand an intimate correspondence and communication between the needs and desires of the settlement population and its representatives, and this one matter should dominate and inform the structure of representative modes of governance.

Given the extreme dependence of life on science and technology, it may be of value, and even a necessity, that the legislature has the capacity to call committees and expertise to deliberate on matters surrounding the creation of laws and points of discussion during its work. However, this capacity might easily become a mechanism for the legislature to extend tentacles of influence throughout the settlement, and vice versa—for other entities to exert direction, such as lobbying, within the legislature. It would be wise to restrict the ability of the legislature to use the excuse of the technological needs of existence in the extraterrestrial environment to expand its powers and reach in this way, perhaps by imposing a numerical limit on the number of independent investigations that it can convene. It might be barred from creating commissions and other bodies under the pretext that it needs external help to enact its programme, as these might wield substantial legislative, executive, and judicial authority and become essentially independent centres of power: miniature governments no less. Commissions may be established to run programmes and projects within the settlement that are of benefit to the population, but their powers must be pruned, and they should never have the capacity to implement rules that amount to independent laws over which they have independent jurisdiction.

It was a noted problem within some of the early state constitutions of the American colonies that the legislative branch tended to dominate matters, leaving the executive moribund. The executive branch of the extraterrestrial settlement must have the power to make decisive moves on an ongoing basis to protect the settlement; in other words, they must have a high degree of independence, even though their actions may be circumscribed by laws and constitutional provisions. For example, the executive may be given veto power over the legislature, with the legislature itself having the capacity to override this veto with some given proportion of support. An independence and vigour of the executive is made necessary by the extremities of the environment in which the society operates and the requirement for effective leadership. However, the executive positions must be subject to regular democratic censure and election, with mechanisms for forceful removal of members of the executive by a given proportion of the legislature for gross extraterrestrial misconduct that threatens the existence of the settlement or its fundamentally democratic culture.

I have not explicated the means by which the highest position in the settlement is to be selected. If a settlement begins in its early stages by the direct election of a president or director from its small population, which seems the most likely arrangement to be realised, I see no reason why this process might not be continued into much larger settlement configurations. If a settlement was of sufficient size to admit of multiple political parties vying for power, the president might instead be selected as the leader of the party that wins the majority of support. Both of these systems of selection, direct and party-based, have worked in representative democracies on Earth; however, the former system would involve less change and disruption to the settlement's political practices, as it grows from an early participatory democracy, than the latter.

A more radical departure from this approach would be to return to the practice enacted by the Roman Republic of electing two leaders (or consuls in their case) that preside over the settlement. This has not been attempted in modern nation states on Earth, but it was conceived as a means to diffuse the potential for an overbearing executive branch. Its obvious weakness is that conflict between these two elected leaders can cause languidness, even paralysis, in the highest echelon of power. It might be noted that the Roman constitution allowed for a temporary dictator to take the reins of government in emergencies, suggesting that in extreme situations they too preferred a single leader of the executive. In the extraterrestrial case, given the range of potentially dangerous situations, a single elected presidential position with the power to act that is subject to checks would therefore seem to be a more prudent design.

We have not considered so far the terms of office for the legislative and executive branches of governance. The electoral period of these individuals must not be so short as to encourage an almost continuous state of intrigue and turmoil, nor should it be so long as to encourage political ossification.[96] Furthermore, to be uncomplicated, the period so decided by experience on Earth must be a round number in the calendar of the environment in which the settlement exists, requiring an approximation to the period on Earth. For example, if we considered an electoral period of four Earth years to meet the political needs just adumbrated, then we might suggest that a simple electoral cycle of two Mars years would make sense on that planet, being equivalent to three Earth years and 279 days; this term is the closest we can get to four Earth years using an easily understood and implemented length of Martian time.

What of the judiciary and its links to the legislature and executive? The first point I wish to suggest is that this branch of government should receive no less attention from the designers of extraterrestrial government than the others. Historically, the legislature has been considered the most troublesome of the three. Locke, Rousseau, the American founding fathers, and many others besides, were greatly exercised by the detail of how to keep the legislature under check. Bucking the trend, Necker is one of the few political philosophers who considered the influence of the executive and its proper control to be the central issue of a successful state. It is almost as if an argument has persisted among political thinkers about the relative effort and discourse that should be expended on achieving success in these two branches and their interconnections, since because one wields the power to act over the population and the other to set the laws, it is apparently self-evident that these two branches should consume our attention.

It surprises me how little importance is placed on the judiciary in the literature of political philosophy, since although it may not exert as much direct control over the functioning of government as the other two, it sets the whole tone and temper of society. A sense of injustice, inequality before the law, and capricious and malign sentencing and punishment; these are the very foundations of social discontent and, ultimately, tumult, not least distrust of government and its power. So a correctly assembled and controlled judiciary whose checks on power are on full display and understood by the public is a matter of singular importance. The public are certainly concerned about how their leaders act and the decisions they take, and they are anxious for the provision of good laws, but above all this, over the long term, their sense of just treatment is the yardstick by which they measure their society and which defines the state of general happiness that exists, irrespective of the short-term political and economic inflections that society experiences.

For this reason, I will later devote a chapter to the matter of the judiciary, the nature of the upholding of extraterrestrial rules and laws, the police force, and extraterrestrial prisons. However, with respect to this immediate discussion on the relationship of the judiciary to the legislature and the executive, we might just note some of the mechanisms on offer to achieve this interconnection of powers. The selection of judges within the settlement might be undertaken by the executive with subsequent review and agreement by the legislature, with the legislature having the power to remove judges responsible for serious misconduct. It is sufficient to point out here that it is essential for the judiciary to be independent and not subject to coercion from the legislature or executive if they are not to become creatures of these other branches, and numerous well-known mechanisms, such as ring-fencing their salaries from being used as a mechanism of coercion by the executive or legislature, can be contrived to realise this end.

The foregoing discussions show that there is an endless vista of speculation available on the possible details of government structure. As the detail is enhanced, so the ideas tend to become more fanciful and naïve. The purpose of this work is not to exhaustively and comprehensively explore all these possibilities and their permutations; my objective is to attempt to draw out the generalities of the challenge for liberty. In that spirit, if there is any matter that I have touched upon in all these schemes, one that seems to be common to any architecture of power that we imagine, and one that should be of special concern, meriting prolonged attention and discussion by extraterrestrial settlers, it is the mechanisms of balance between the branches of government. The separation of powers is one blunt means to dissolve the potential of autocracy, but it is the relationship between those now separate branches and the checks on each other's powers that sets the tone for the entire edifice of freedom. Weak restraints can lead to one branch dominating or to a structural laxity in the whole government that gives way to tyranny. Overly strong checks can cause seizure in the functioning of government. This difficulty is common to any proposed structure and scheme of government architecture; it is universal to any design that the settlers should eventually decide to alight upon.

The matter of how powers are separated and how they check each other is of particular interest because the space environment encourages expansion and encroachment of any branch of governance. The executive will not wish to be seen to be ineffectual in protecting the people they govern under such extraordinary environmental extremes, so they will be in constant search of new powers. Seeking to avoid accusations of injustice or arbitrary behaviour, they will be tempted to encourage the legislature to create new rules and laws under which they can carry out their functions. In turn, the legislature, not wanting to suffer the opprobrium of having created insufficient laws to protect citizens, will seek to expand the number and remit of rules, regulations, and laws into all aspects of human life through quite well-meaning intentions—and it will seek to ensure that the executive implements these laws which they have gone to significant effort to create. The judiciary, for its part, will be constantly on edge, thinking that it has failed to implement laws that protect people. Thus, they will expand their power and action, but at the same time, they will seek to encourage the legislature to provide them with the means and the executive with the latitude to act. Hence, we can see that the extremities of the space environment provide the excuse for every branch of government not merely to explore the means to expand its power, but also to encroach on the others.

It is not clear that lessons on Earth can be transplanted without modification into space. The novelty of the physical conditions and exigent difficulties that might emerge likely require these checks and balances to be crafted and honed as the reality of governance in the extraterrestrial condition becomes apparent. Extraterrestrial constitutional conventions and regular and intense debate on the political provisions generated in the early years of the establishment of a settlement will allow for refinement and the creation of an order that commands the greatest proportion of support.

One way in which we, and extraterrestrial settlers, might ascertain the legitimacy of any governance structure devised, and continue to review it, is to imagine a Rawlsian veil of ignorance. Milligan has provided a cogent discussion of how such a mechanism might be used for deliberating constitutions beyond Earth.[97] Given the task of devising an institutional structure, would the chosen governance structure recommend itself to you if you had no knowledge of where in that society you yourself would end up? These mechanisms of thought are potentially useful in devising optimised arrangements that have some chance of creating organisations of governance that arouse the support of the widest franchise possible. Furthermore, in that they ask participants to imagine real situations, they may be less liable to fall prey to idealism; they may at least align aspirations with expected real outcomes when those using this Rawlsian device consider the worst-case scenarios into which they themselves might be deposited.[98] In practice, the application of this procedure may be relatively uncomplicated in that all that is being asked of people is that they consider the provisions within constitutions, and other rules, in the context of an assumption that they might be required to live under them.

In this section, what I have sought to illustrate to the reader is that societies can likely be assembled in space in which the conditions for liberty can be deliberately enhanced and that lessons from Earth may yield much of use in devising the structures of government that maximise the chance of that outcome. Furthermore, when these structures are eventually constructed, there should be no end state to the process of political development. 'Every human institution must stand under constant review', as Niebuhr observed.[99]

An essentially classless extraterrestrial population can facilitate the emergence of a successful social order of a free republic

Despite the dangers that I have drawn attention to within the quintessentially revolutionary character of a new society in space, there is an aspect of this paucity of precedent that could, if realised and nurtured, enhance the chances that the social order and the separation of powers that I have discussed can be made to settle into forms conducive to an unautocratic social order.

A historical complication in realising the separation of powers in emerging governments on Earth has always been the problem of mapping onto the segments of government—that is, the executive, legislative, and judiciary—existing classes of people within that society. This multifactorial problem introduces great complication, because now, not only must the formal branches of government be made to balance each other without any one of them gaining destructive domination over the others, but the classes of society must also be managed within this structure so as to ensure than none of them become overbearing to each other or use the branches of government to that end. For example, in the case of England, the legislative was constructed such that broadly the lower house of the bicameral system, the House of Commons, was instituted from among the bulk of the people and the upper house, the House of Lords, provided voice to the aristocracy, allowing both to check each other and to ensure that both classes had active participation in the legislature and neither would feel spurned by the other.

The dilemma with these arrangements is that they depend upon a propitious set of historical circumstances for the pieces of the jigsaw to fall into place without rupture. Diversions into entertaining but questionable 'what if' historical scenarios are not necessarily helpful, but it is worth noticing the many ways that England might have failed to achieve a balanced system of government. If Oliver Cromwell had been marginally less brazen, but still dictatorial, perhaps England would have settled into a long-surviving autocratic republic. If the Glorious Revolution had not led to a successful shackling of the monarchy and the assertion of parliamentary supremacy, a long-lived absolute monarchy might have driven English liberties into a monarchic suppression of all branches of government which could have endured (motivated by the monarchical fear of another regicidal decapitation)—and so on, speculation ad infinitum. The point is that liberty is a fragile thing and made more fickle, unpredictable, and prone to death when the complications of class structure and other social fractures are thrown into the arena of an already highly complex problem of creating checks and balances between segments and branches of government.

A vigilant population jealous of its freedoms can inexorably lead a society towards a state where those freedoms are given wholesome expression, even when powerful class structures exist, as the English have, on the whole, managed to do for several hundred years. However, surely any society that has the capacity to assemble itself from scratch would wish to minimise the hard work and the potential number of perils and pitfalls that might lie in the way of that journey?

The relevance of this history to the extraterrestrial case is that provided a society is not initially assembled from people who strongly associate with disparate classes, the problem of creating mutually supporting structures of government without these secondary internal tensions can be avoided; the matrix of potential institutional discontents can be simplified and the combinations of potential conflicts minimised. More to the point, the instigators of the first permanent settlements should deliberately set out to avoid the presence of potentially antagonistic class-like stresses within a founder population; this is not merely to achieve general social concord, which will no doubt be self-evident to them, but for the purposes of enhancing the later capacity for that society to create more complex governance arrangements, including the separation of government powers, in which latent social divisions will not find destabilising expression. In this sense, the revolutionary impulses of a new society in which classes have not had time to congeal and solidify can themselves become the steadying influence in testing the sharing and distribution of power without arousing the covetousness for power between classes that exist in older societies possessed of such well-formed social distinctions.

Access to information is essential to promote liberty in extremes beyond Earth

Throughout all the foregoing discussions there has been a matter that is essential for the success of society, which until now I have deferred from discussing directly: to create effective systems of free deliberation, citizens must have access to sizeable and meaningful information about their situation.

Any settlement in an extraterrestrial location is physically constructed, and continues to be developed, around a set of technical decisions. Such a statement would apply to the building of a house on Earth, but the decisions taken in the extraterrestrial environment will often revolve around the potential for instantaneous death in the external environment; there is no 'buffer' to existence in the natural environment.

One consequence of this reality is the cardinal importance of being able to freely access information about the apparatus that sustains life continuously against the backdrop of the deadly threat. I contend that this lies at the heart of any successful free polity in space to a greater degree than on Earth. For any individual to be able to understand a decision, either to challenge it or simply to displace a conspiracy theory, as much information as possible about the technical and social aspects of the settlement must be on full display. It is likely to be the case that individuals may not necessarily have the technical knowledge to judge certain decisions in the settlement. However, the culture of deliberately making available information concerning all the different aspects of running it will go a long way to achieving a society in which the people feel involved and will lessen the suspicions that attend any given action. In today's lexicon, we could say that the society should strive for that much over-used word: transparency. Beyond Earth, this word takes on solid meaning.

There are a range of ways in which access to information might be achieved. In a small settlement, it may be realistic to train as many of the personnel as possible in the numerous structural and life-support systems. As a settlement grows in size to tens or more, where universal knowledge of the various elements of the infrastructure is impractical and unlikely, we might imagine that deliberations on oxygen, food, water, and power systems would be made fully public and that decisions concerning the manufacture, distribution, and ownership of these most vital human requirements would have to be carried out by groups and committees that are subject to public dissemination of their considerations. This suggestion does not imply that a utopian transparency is possible—individuals can always meet 'off-record' to make decisions —but the intent and demand that the bulk of these deliberations can be seen by the population may be an important culture to construct.

Of special significance are decisions that involve organisational structures. Cabal and monopolistic attempts to deliberately secure control of vital commodities such as oxygen represent the most pronounced threat to the community, since once a monopoly on oxygen supply is coercively obtained by a group of people, it may be difficult to wrestle it back, except through the extremely dangerous mechanism of revolution.[100] The community will need to be alert to these political and economic developments, and an approach to increasing awareness of these is to make the informational structure of the settlement as leaky as possible, ensuring that as many as, even if not all of, the decision-making processes surrounding vital commodities are open to public censure and insight.

The means by which this information can be provided is by an unusually rapacious culture of the provision of meeting minutes, composition of committee structures, transcripts of essential gatherings, and so forth through electronic media for anyone to access, without the requirement for formal freedom-of-information bureaucracy. One might summarise by observing that an extraterrestrial society could, with deliberative intent, generate a culture where administrators and the population at large see it as something of an honour to propagate and make information widely available on vital production and processing capabilities, with the legal system and rules to sustain that.[101]

An effective means to improve the dissemination of information is to encourage spirited and vigorous discussion through news outlets and other journalistic contrivances—essentially, a nascent extraterrestrial free press. In the very early stages of a settlement, these outlets might include amateur periodicals in the style of Antarctic station newspapers, political pamphlets espousing varied political positions,[102] and debates on a variety of multimedia; as a settlement grows in extent, it may expand to a fuller range of these outlets. If gentle rivalry between these organs for stories exists, in the same manner as news agencies on Earth, investigative journalists will be encouraged to find their 'scoop', sharpening the alertness to venal activities. These journalistic proclivities will not be popular, but a test of those running an extraterrestrial settlement is whether they are suppressed or encouraged. For an extraterrestrial settlement, the latter must be the case, because it is self-evident that the discomfort a free press can cause is more than made up for by the window it provides on processes that may not be in the interests of the wider polity.[103]

The capacity for individuals to freely own and operate technology that can receive information from other people, or indeed any technology of information transfer, should be an unassailable principle of free governance, since without it the foregoing suggestions are rendered sterile. The policy implication in the extraterrestrial case is the avoidance of all regulations, fees, and other mechanisms that would stymie or slow the capacity of individuals to own technology or gain information that may better their knowledge of any subject for which they feel a need to master, particularly when it concerns the running of a settlement.

In essence, to an extent not witnessed on Earth, the extraterrestrial free society, to achieve a peaceful trust and contentment among people confronted by disabling conditions, must develop mechanisms and cultures that diminish any suspicions of cabal, cronyism, or monopolistic control inconducive to the common good. The people must be confident that where such inevitable human behaviour emerges, access to extensive information can be gained with such ease that it will quickly be detected and remedied by society, and that within this openness, the safety of their lives, and the survival of their liberties, is assured.

Communism is a system unfit for humans on Earth and beyond

The reader may have noticed that I have not proposed as a governance structure a truly socialist commune where private property is eradicated and the whole group of people live in a form of communist interdependence.[104] It might even seem that the lethality of space environments would make such arrangements a logical way to deal with conditions in which a rampant individualism could jeopardise the society.

Furthermore, individualism aside, we might note that the factors which I have drawn attention to as most attractive to a tyrant, particularly oxygen production, can be calculated exactly and thus apportioned without the possibility for arbitrary use. The maxim popularised by Karl Marx, 'From each according to his ability; to each according to his needs',[105] has a rhetorical appeal in space precisely because the needs people have for oxygen, water, food, and power can be determined scientifically. Lenin described Marx's popular maxim on needs as a 'higher aim',[106] and it can be sold in this way with relative ease on the Moon or Mars, the aim here being to design with engineering accuracy and precision a system of provision that meets every individual need with exactness. Surely, here at last, ranged under instant lethality yet with the requirements for existence known with scientific certitude, we have discovered a place where the centrally planned socialist society can be achieved? It is an immensely attractive conclusion. However, although extraterrestrial needs may fit the Marxian credo more exactly, I think there are wider failings with the idea.

The failure of the communist modes of political and economic activity on Earth is often explained away by the suggestion that humans are not good enough for communism; their selfish interests and individualised versions of life grate against the ideals that communism demands. Concomitantly, where failure does occur, this is therefore rebuffed with the view that it was not proper communism. Of course, I think this is the wrong way around. Communism and strong versions of socialism are systems inadequate for humans, which because of their cognitive capacities, their complex emotions, needs, and wants, all have individualised versions of the good life. Any political and economic system that demands obedience to a central plan may be suitable for an ant colony, but it is utterly inadequate to meet the never-ending and continuous changes in individual outlooks and economic demands of the human mind.

Contrary to being ideals which humans have a hard time matching up to, communist visions of human existence are unworthy of the expansive creative capacities of humans. Although one might argue that liberal systems of government, particularly ones aligned along capitalistic economic models, demand allegiance of a kind, in general they demand no fealty to a monolithic central plan as such; the system admits of far more individual divergence (hence my confidence in writing this book and critiquing aspects and institutions of the society in which I live without worrying about what others think of it). In a free society, properly so called, it is possible to attempt to convince one's fellow citizens of radically different economic and political arrangements to those in existence, a tolerance which will always be threatened in a communistic system pursuing a unified vision of how all citizens should interact.

Any system that is established on a narrowly defined ideology, communist, religious, or otherwise, can never tolerate dissent outside that narrow corridor, since by definition it threatens the survival of the ideology. I will not be so naïve as to claim that liberal systems of government do not have limits to dissent. If you establish a communist party in such a system and actively propose the overthrow of the state, you will surely attract attention from the state and fellow citizens, potentially of an unpleasant and vituperative kind. But without an unyielding ideology, such systems of governance do, in general, allow for a wider corridor of intellectual manoeuvre. Furthermore, I will not hesitate to point out that this freedom is denied to the impecunious, and there are plenty of other weaknesses in democracies and capitalism, but there is no counterclaim here that liberal democratic systems are utopias.

A system of private property, and with it the associated culture of possessing a sphere of private dominion and jurisdiction, is essential to developing a powerful capacity for self-expression and creativity. This energised sense of individuality need not compromise the rest of the group or endanger it, but it may aid in bringing forward a variety of characters who can offer flexible types of leadership, expertise, and ideas of benefit to a society at the physical extremes of existence.[107] I think a relevant analogy here is the ancient Greek societies of Sparta and Athens. The former was constructed as a military society of uniformity, the latter on relatively liberal principles (at least by ancient standards).[108] Both societies succeeded in surviving in the Mediterranean environment, but we would have little hesitation in recognising the vastly greater contribution made by the latter in philosophy, medicine, biology, astronomy, arts, and culture. No matter what the social complexities of the reasons for this blossoming of the human intellectual condition in ancient Greece might have been, it must surely have been the case that the freedom to express these ideas in the first place, unrestricted by military coercion and servility, was a necessary precondition.

Even if one could construct an argument that a communist system would function in an extraterrestrial environment, even succeed, I suspect that it is less likely to spawn an efflorescence of human creative capacity and happiness in the long term compared to a society where the individual has a more expansive domain of self-expression, property ownership, and non-interference.[109]

There are other factors that bolster the case against a communist system. The first is the inflexibility of a system rejecting private ownership. An individual moving from such a system into one where private property is allowed can obtain as much or little property as they wish, but in the inverse situation, the individual must give up everything to be admitted into the society, so as a system to be propagated across the cosmos, a communist commune is liable to be less flexible in the way in which people move in and out of it compared to one where settlements with private property admit of a whole range of ownership paradigms.

Second, despite good intentions, the lack of a private domain of property of any sort leaves an individual utterly dependent on the collective organisation and goodwill to get anything that they need. Such a system may well engender happiness when the leaders are benevolent, but any malfeasance from leaders or even citizens leaves a person potentially more exposed to coercion when they own nothing or very little by political design.

Third, the need for a philosophy that creates a collective commune requires that everyone accept that philosophy and submit themselves to the commune.[110] Although one could say the same thing for a system of private property or democratic government, these arrangements are far more forgiving and open to personal interpretation of the good life than systems where complete obeisance to the collective is required. These latter systems tend to be less expansive in their visions of human liberty and its kaleidoscopic expressions.

Fourth, fearing that the whole edifice of the communistic state will be brought into disrepute from failed calculations, an inadequate central plan, or even just inefficiency and incompetence from officials that may be given expression in quotas and productive capacities that are not met, failings which in the case of oxygen production, for instance, could be cataclysmic, the governing powers will always be wary of information that is to their detriment. Freedom of access to information, so essential to involve the population in society effectively, will be under constant erosion as the central planners seek to project their efficient management, hide failings, and distract attention from information that may attract criticism. The ultimate effect of this venality may be catastrophically fatal.

These failings of states are familiar on Earth, but we should realise that some of them are intensified in space. In particular, the pressure to correctly centrally plan commodities whose want could cause an existential emergency will cause a particularly rapacious desire to control the flow of information and cover up mistakes; this culture in itself becomes a serious threat to existence.

There is a long history of attempts to construct socialistic utopias on small scales analogous to the extraterrestrial case. I have discussed elsewhere the kibbutzim, but examples abounded in the nineteenth century during a period of fervent experimentation in communal living. George Ripley's Brook Farm, a communal experiment in living established in Massachusetts during the 1840s, is a case in point. Starting with an egalitarianism that espoused freedom of speech, the failure to inculcate an order into the community, encouraged by the resentment felt by some for their equal pay alongside others who failed to take responsibility, led to a more intransigent socialist structure inspired by Charles Fourier.[111] It eventually dissolved. Brook Farm was not unusual. Its failing was not merely caused by an attempt to form a nebulous utopia without a credible plan for reaching that state (and a contributory motivation for my own vision of building anti-dystopias instead); in fact, the communal structure that required obedience to a central plan was unyielding to the challenges of the community as its needs developed.

Not all of these valiant attempts at human social reconstruction have failed in the sense that they collapsed. Shaker communities lasted for many decades. Yet it is true to say that most of those that lasted for lifetimes either had dogmatic and highly influential, even cult-like, leaders or powerful religious monopolisation.

The communal experiments attempted in the past, and particularly those dreamt by social reformers with a socialist hue in the nineteenth century, provide a wealth of ideas, lessons, and insights. We might recognise that less extreme versions of utopias, such as those furnished by Josiah Warren and Robert Owen, led to concepts of workers' cooperatives that may have much to teach us about how to meld the need for individual freedom with the more collectivist spirit required in extraterrestrial extremes. However, I think it is true to say that almost exclusively they provide lessons about how not to create communities.[112] In the extremities of space, where considerable resources must be mobilised to even establish a settlement in the first place and there are more profound difficulties in acquiring resources and organising exit than in the fields of Massachusetts, a lunar or Martian Brook Farm is inviting considerable disaster. On the other hand, one should not admonish the idealistic and well-meaning ambitions of those who committed their finances and personal effort to building these societies. In fact, they were laudable, hopeful visionaries, and in the absence of such experiments we would not now have the insights that we do. One might observe that, unknown to them, the experiments they attempted and ultimately failed in on Earth may one day save the lives of many people on the space frontier if we heed what they discovered. That experiences gathered in an isolated farm in West Roxbury in the 1840s would one day inspire people to discuss how best to manage a settlement on Mars, I expect they would have found astonishing.

An extraterrestrial governance structure can perform many functions

At this juncture, it is worth clarifying what particular roles a government structure might assume. I have discussed some of the preferred structures of governance, but have not formalised exactly what will be the domain of coordinated action by this structure of governance, whether within small participatory democratic settlements in which everyone partakes or in representative bodies of a larger settlement. These are the operations that will be overseen by the governance structures that are not left to individual effort, private corporations, and other forms of voluntary association. Klosko[113] provides a useful non-exhaustive list of the role of the state, and I take that lead to suggest activities of the extraterrestrial state.

Listing roles of a governing structure, or state, is a contentious one, in that depending on the collective or libertarian slant of the reader, there will be some disagreement about the properly circumscribed powers, just as there is on Earth. In particular, the *degree* of private involvement in the provision of certain functions is, I think, difficult to define. Aside from the overarching political and economic philosophy of those running the settlement, it depends on the practical realities of many unpredictable interactions between individuals and organisations. More private activity in one activity can encourage more or less similar approaches elsewhere in a political system or economy. However, the extraterrestrial environment does require a set of tasks to be completed, some of which are unique, or at least particularly relevant, to that environment. This does allow us to state what some of the responsibilities of the extraterrestrial governing structures might be:

- Implement the democratically agreed decisions of the settlement (a form of executive branch).
- Construct and oversee a legal system, in terms of both setting laws (a form of legislature) and implementing them (through a judiciary, including looking after any prison facility). For a small settlement, these functions may be highly informal and may take the form of 'rules' rather than 'laws' as such.
- Coordinate a secure environment by enforcing certain rules, particularly those that pertain to situations associated with a high risk, such as depressurisation and fire.
- Establish and oversee operating and material standards, and their associated rules, for habitats and other pressurised spaces across a settlement, particularly those that pertain to matters such as radiation protection, dust mitigation, etc.
- Set and oversee safety procedures and standards for oxygen, water, food, and power production.
- Cooperate with all entities to ensure a minimal level of oxygen, water, food, and power output is achieved with sufficient redundancy of production.
- Assist with, and encourage, the implementation of educational opportunities for all age groups.

- Coordinate with all entities to ensure the implementation of a 'welfare system' to assist the temporarily or permanently indigent, incapacitated, or elderly.
- Coordinate and implement forms of healthcare in a settlement.
- Facilitate and encourage scientific and artistic activities by creating spaces where these can occur.
- Facilitate voluntary association by working with all entities in the settlement to draw up the rules by which societies and other organisations will operate, and aid them, where possible, in coming into being.

There may be others, but we can see at once that for a society established on the blank canvas of an extraterrestrial landscape, or in the free plateau of space, there is an immense amount of work to do. Even if one were to relegate many, or most, of these activities to private corporations, we can see that there are many activities that would still require some form of coordination.

I provide this list to illustrate that the development of systems of governance in space that are accountable, and in which power is checked, is an essential precondition to creating a governance structure that can contribute to all these activities, or even just assist them in the private sphere. We can see that if a government administration is well designed, then there is no need to fear its capacity; rather, its potential should be utilised to coordinate and complete many tasks.

Voluntary association is essential to disperse extraterrestrial power

As the preceding has shown, there is a very considerable amount of work to do to achieve survival and success in the extremities of an extraterrestrial environment. Although all of this could be delegated to some commanding authority, there is sense in creating a range of other organisations that can help to meet these diverse challenges and needs. Beyond these practical considerations in dividing labour, a concern I have raised at multiple points so far is the tendency towards an all-encompassing centralisation in extraterrestrial power; the check on central control can be facilitated by encouraging to the highest degree societies and associations that are separate from the groupings that constitute the main branches of governance. I mean by this a variety of intersecting, overlapping interests in the form of voluntary associations, which could range from something as trivial as a settlement reading club to a hobby engineering group. Other examples might be associations of interior designers, spacesuit design and engineering groups, art clubs, and writing societies. We can also refer to these as 'intermediate' organisations, since they sit between individuals and the highest governing structures.

The purpose of these intermediate groups is to enable conversation about the settlement across many separate centres of power representing forces that are independent from those at the centre of the settlement.[114] These organisations stand, if you like, between the amorphous whole of the population and the governance structures and give voice to the varied interests and views of the populace.[115] They act to diversify and deepen the range of discussions about the predicament of the settlement, thus diluting the power of the structures of governance and at the same time increasing the range of ideas and solutions that are on offer for solving the settlement's challenges against that inevitable tendency to conformity. They provide a physical conduit to discuss issues related to the settlement and to challenge ideas. Furthermore, they can act to slow the implementation of central directives and policies either through direct pushback or by acting as a diffusive barrier to the implementation of directives at the local level. Although this may be undesirable in an emergency, it can act to provide a temporal space for the emergence of valuable dissent.

Aside from these functions at the large scale, voluntary associations provide the context in which individuals learn and practise the manners and habitats of democracy; by this I mean expressing views and arguments in a coherent way, developing forbearance towards others and their ideas, engaging in electoral processes, observing and appreciating the shifting sands of opinion, and learning the mechanisms and processes of associative activities in their widest meaning. If the democratic process is only rarely and infrequently expressed by elections at the highest level, people lose that democratic culture and natural insight. When they do express it, they may do so clumsily, allowing passions and irrational fears and demands to take hold. A continuous process of deliberation across a great variety of activities and social interactions maintains the health of the democratic outlook, not only making it more effective and calmer when applied to the governing authorities themselves but also generating a continuous spirit of free democratic conversation at all times and across the whole population.

The ferment of opinions that emerge from voluntary association can do much to stir up that threat of a flat conformity to which I alluded in Chapter 2. Instead of agitating the waters of public opinion irregularly during elections for the governing authorities, voluntary associations are, if you will, a stick that constantly keeps the waters in motion, allowing individuals to contribute new ideas and to voice disagreements without standing out from the crowd to the extent that they are given the side-eye by the community as a whole. Breaking conformity is a positive feedback process; the more the culture allows for dissenting views, the easier it is for individuals to engage in this dissent without fear of attracting damaging forms of attention. However, this circular process can be easily quashed by a majority culture, and it is necessary to make all efforts to keep the waters of public opinion constantly in flux. Encouraging voluntary association is one mechanism to achieve this end.

As with all matters to do with liberty, let us not be utopian either. Friction between governing power and these 'intermediate' organisations can lead to illiberalism as organisations seek to assert dominance.[116] Nevertheless, all things considered, a society in which people are free to choose purposes and ends and to combine with others to realise those plans is likely to be more conducive to free thought, and thus to creative solutions to society's challenges, than one in which voluntary association is suppressed. It also follows that to allow the effective expression of these groupings, there should be no law or regulation that prohibits people from participating or puts limits on the number of people that can associate to realise these ends.

Quite apart from diffusing central power, voluntary association of any kind, when expressed effectively across a wide range of human actions and activities, has the ability to water down the influence of influential vested interests or organisations. The encouragement of diverse associations and their formation is proposed as an explanation for the emergence of religious pluralism in the US[117] and the dilution of the possibility of an all-embracing monotheism early in the nation's history. Whether or not we agree with those arguments, it must surely be a logically correct assumption that if a society has a plurality of voluntarily formed groups, then the diversity of thought is greater and the power of single institutions and their capacity to dominate thinking in society can be more easily lessened.

We could consider a particular historic situation. Earlier, I drew attention to the dictatorial regime established on the Scottish island of St Kilda by the nineteenth- century minister on that island. We should recall that this occurred in a community that possessed a participatory democracy, illustrating that illiberal behaviour can live side by side with democracy. How did this occur? The democratic deliberations of the community did not prevent a single individual from controlling the cultural life of the island. In this historic case, religious beliefs structured the spiritual lives of the islanders under the despotic hand of a single individual, while practical matters of survival and living continued within a democratic framework.

The antidote to St Kilda's problem is to ensure that the cultural life of a community is not directed under a single monolithic vision but that overlapping voluntary associations touching on many different cultural interests are created such that no single individual can easily dominate. However, like constitutions, voluntary association is not a singular remedy for tyrannous social ills. If St Kilda had been subject to more movement of people in and out of the settlement, the behaviour of the minister might have become better known to outsiders on the mainland, who perhaps could have acted to remove him or enter the settlement to reduce his influence; the isolated islanders received no meaningful redirection from the outside. Therefore, one additional lesson to be learned for the extraterrestrial case is that extreme isolation should be avoided. I will return later to the matter of how we might deliberately engineer the capacity for the free movement of people in and out of extraterrestrial settlements to mitigate this St Kilda-like isolation of populations. Nevertheless, in the context of my discussion here, the point is that voluntary association should be implemented alongside other social mechanisms. It is part of a mosaic of methods, including free movement, that together can reinforce the strength of community bonds yet at the same time dilute centralised human power.

Faced with the deadliness of the extraterrestrial environment, there is a strong temptation to see intermediary organisations of any kind as a threat to the unitary decisions that may be necessary to prosper under the conditions at hand. A society of atomised individuals herded under a single dictate and led in a clear direction would seem to have a chance of more success than a society in which individuals are continuously being pulled this way and that by differing associations, societies, and groupings, all of which have varied agendas and interests to represent. Although I think that in situations of extreme danger this is a legitimate concern, the general ethos of society should seek to rebuff this conclusion. In the longer term, the survival of liberty and the avoidance of an uncompromising power structure requires these intermediary organisations, however irritating they may be to the purists who seek clarity and simplicity in social decision-making.

The involvement of the population in the polity by lot

Aside from voluntary association, other mechanisms might be deliberately instituted to both increase the involvement of the population in the political affairs of the settlement and enhance the multidirectional conversations occurring across different groups within the population. An example concept could be the formation of committees, populated by lot, whose task is to oversee and discuss all matters associated with the key requirements of oxygen, water, food, and power. Charged with taking part in meetings of the governance structures, visiting facilities, prisons, etc., their purpose is to write regular reports on the state of these capacities for their fellow citizens, while having the power to raise requests and any apprehensions about these vital resources. They are essentially independent groups of people who provide secondary oversight and opinions on the workings of the key areas of production that are likely to fall prey to centralised or tyrannical control. They will help accomplish that need for the dissemination of information about the settlement which I have already discussed. The period of such committee work could be equivalent to six months to one Earth year, and the economic and community arrangements in the settlement might be made with this requirement in mind, not unlike jury duty in some nations on Earth.

The mechanism of selecting such committees by lot, or sortition, is an inspiration from ancient Athens[118] and Italian city states,[119] and I think it has three aspects to recommend it. First, it ensures the involvement of the whole populace, thus exposing them to the workings of their society with active political involvement. Even in the event of a languor towards politics, this responsibility provides a duty of involvement. Second, by knowing that a time will come to serve, the system encourages people to take an interest in the arrangements of the settlement because no one wishes to come up short. In anticipation of the call by lot, many people might take a pre-emptive interest in their political and economic situation to ensure some level of competence in the tasks eventually asked of them. Third, this obligatory involvement achieves a measure of accountability in the democratic process.[120] Individuals must take part in the political process, and in so doing they will come under scrutiny from their fellow citizens and they themselves will find themselves in positions in which they must make judgements on the work of others. Through this mechanism, the government is not only a set of servants doing the bidding of a population otherwise detached, but the people and the governing structures become accountable to each other. This characteristic is particularly germane to the extremities of space with respect to liberty. By randomly placing people in parts of the settlement where they have some temporary oversight and responsibility (although not necessarily complete responsibility for many technical activities, since if they do not have the necessary skills that would be to court mishaps), the population may gain the education and knowledge to play a fuller and more informed role in the polity in general. The mechanism of lot may therefore enliven an understanding of the wider workings of the settlement and subdue ignorance that is so beneficial to the tyrant.

As with the ancient Athenian council of five hundred (βουλή), which was the full-time group charged with running the daily affairs of the city and setting the agenda for the assembly or ekklesia (ἐκκλησία), the selection of individuals by lot could even be applied to some parts of the executive and legislative functions at the core of the settlement's governance structures. I see no reason why an extraterrestrial settlement could not profitably experiment with both election and sortition for various of the wide range of political tasks in society, settling on its preferred balance in the means of democratic selection. The mechanism of lot, although of particular value in direct democracies, could be continued in a wide variety of tasks, even at enlarged scales in more representative modes of democracy, as a means to ensure public participation in the activities of the settlement.

Alongside wide public involvement, mechanisms might be provided for a healthy turnover of officials to reduce the opportunities for the ossification of networks of power. In a small settlement, the choice of people may be few and it may be difficult, if not impossible, to achieve a regular shifting of personnel. Yet we could deliberately train and educate people within a settlement, even at an early stage, to have the capacity to take on different jobs, thus allowing for a fluidity in their movement. Peripatetic mechanisms, such as fixed periods in certain health and safety tasks, or even democratic voting processes for senior officials involved in health and safety, may be one way to keep a population in some flux and continuously wary and alert to these networks of officials charged with ensuring the safety of the settlement. This approach to democratic involvement, it should be noted, would have the side-effect of enhancing redundancy in the settlement by expanding the training of citizens to be capable of undertaking a wide variety of tasks.

The settlement should desist from titles and formal systems of recognition

To reduce the potential for arbitrary rule and strengthen the influence of these interconnected webs of voluntary association, as well as to protect the culture of any society, I think it is enormously important for governing authorities to desist from systems of awards, honours, and titles. Article 1 of the US Constitution, establishing the limits of the legislative branch of government, states: 'No Title of Nobility shall be granted by the United States' (Section 9). This provision obviates hereditary titles. In the extraterrestrial case, it will be fiercely tempting for a governing power to concoct a range of awards and honours to show its approval of people that support its agenda, accolades which may or may not be hereditary.

The flaw with a formalised system of honours or titles of any kind is twofold. First, no organisation can possess full information on any individual to assess their broader social worthiness. In the same way the difficulties inherent in economic centralisation are caused by a lack of information, the 'central planning' of honours and titles is always encumbered with the possibility that the awards will be given to people who later turn out to be unsuitable. Simply rescinding the award in the event of serious social transgressions hardly addresses the problem, since the error is in a system that attempts to make centralised judgements on people in the first place. Second, it has the potential to exert an unhealthy influence on society, essentially generating a form of institutionalised deference to the state and to the people and bodies from whom these awards are distributed. People can seek to be part of the system of awards and titles for the sake of being titled, with the work and achievement that underpins it becoming of secondary importance. The system has a tendency to generate ambiguity in the motives behind human action, whether they are driven by a desire to accomplish a task or to receive state recognition; a corollary is the tendency to weaken the vivacity and energy with which people are willing to critique the fundamental structure of the society in which they live.

If there is to be any form of honour system in a society, then it is best limited to the sorts of awards associated with specialist organisations that are free to recognise whom they wish. At the higher level, and by this I mean a person's recognition more widely in society at the level of governing bodies, this should be left to an organic process whereby a person's respect and approbation by society as a whole is determined by the many and varied interactions they have in society and the ineffable and nebulous view that emerges about a person from these activities. The state or its equivalent should seek no part in this process.

The achievement of this neutrality is not a complex matter; the US Constitution established it in a mere 11 words. In the extraterrestrial case, a constitution might state that 'No forms of titles or honours shall be granted by the governing authority.' Despite the triviality of this mechanism, the barring of centrally planned recognition is immensely important in weakening the state's involvement in the artificial organisation of society, especially in the extraterrestrial case, where the governing bodies may seek allegiances to consolidate a hold over an isolated group of people confined in extreme conditions.

Economic freedom is essential alongside extraterrestrial political freedom

Although economic activity is an irrefragably overlapping part of the political sphere, it is worth considering some specific problems and solutions to the economic challenge. Economic encouragement to tyranny is not disconnected from the political sphere in that the loss of freedom in political circles can lead to a loss of economic freedoms, for example in the choices that employees have in their vocations. A collapse of the economic system into special interests can itself have a converse devitalising effect on the health of political freedoms. Many of the remedies for economically encouraged tyranny therefore have necessary and unsurprising links to the political mechanisms I have adumbrated. Economic and political liberty are conjoined twins.

Perhaps the clearest channel to economic despotism in space is the conditions brought on by isolation. Physical isolation in space need not lead to economic tyranny as such, but we have imagined a number of ways in the previous chapter in which it might do. How are these possibilities to be forestalled?

A sure way to alleviate many of the economic blights to liberty that have been identified is to make it easier for people and goods and capital to flow between settlements and within them. Much of this has to do with using engineering to overcome the restrictions on movement that the extremities of space cause and which lead to isolation, the cause of some of these malignancies in society. In a later chapter, I will have more to say about 'freedom engineering' and how we might use practical engineering means to overcome some of the naturally imposed causes of isolation, examples being the mass production of reliable spaceships that enhance the potential for movement.

Whether we face a situation of economic centralisation and coercion caused by a government or a private corporation, I suspect that one of the best ways to prevent economics from becoming a facilitator of tyranny is to encourage a diversity of operations and providers.[121] A single provider of oxygen machines can have many dramatic effects on the economic conditions within that specific sphere. The users of the machines come under the influence of that single organisation, but then so too may the people providing the spare parts and the servicing. Even if we have taken care to prevent the centralisation of oxygen production into a single machine, it does little good for everyone to own their own machines if the upkeep and servicing of these machines still depends on a single entity. We would be wise to design oxygen machines that can accept spare parts from many different economic outfits and to ensure that these enterprises actually exist and operate in a given settlement, so that the choice of a provider of spares and of the machines themselves is made real and is not a mere abstract optimum economic desire.

In a small, isolated settlement in an extraterrestrial environment, a fully fledged market where prices fluctuate fluidly as many people buy and sell oxygen machines of various kinds may be idealistic. The difficulty in building factories, extracting raw materials to supply them, and creating the supply chains to make them viable might be extraordinarily challenging, and we could imagine that a high critical mass in manufacturing capability and people would be needed to create the sort of multifarious dynamic markets we find in nation states on Earth.

Yet, despite those concerns, one should not underestimate how human ingenuity can generate markets with little planning and with few products and in locations with confined populations. Radford's seminal paper[122] on the economics of a prisoner-of-war camp in the Second World War provides a particularly lucid example of the emergence of markets within a small population of people under conditions of extremity. Drawing on his experience of Oflag prison camps (camps for officers) with population sizes between 1,200 and 2,500 people but themselves split into companies of about 200 people, he explains how a currency system in the form of cigarettes spontaneously emerged and was used to produce a miniature economy in a whole range of products. Speculation and banking even took root. Radford observed: 'But the essential interest lies in the universality and spontaneity of this economic life; it came into existence not by conscious imitation but as a response to the immediate needs and circumstances. Any similarity between prison organisation and outside organisation arises from similar stimuli evoking similar responses.' This system was not motivated by negative selfish interests. Radford explained: '"Goodwill" developed into trading as a more equitable means of maximising individual satisfaction.' Equally significantly for the extraterrestrial case, at least in the early stages of a settlement, he observed: 'It is thus to be seen that a market came into existence without labour or production.'

Radford's exposition is highly intriguing because no deliberate attempt was made to create this system, as might be encouraged in the extraterrestrial case, yet it still thrived. The very pressure of physical extremes and the threat of the want of vital resources in the extraterrestrial case, as in the prisoner-of-war camp, is likely to be the catalyst for vigorous market mechanisms, not their nemesis. Without much planning, we can expect that an extraterrestrial invisible hand[123] will emerge beyond Earth, the broader economic activity guided by individuals each acting in their own interests, provided the social environment allows for it.[124] I take it to be essential that for these economic drives and impulses to appear and proliferate, free rein must be given to their emergence. This implies that restrictions on trade in its many possible forms should be minimised.

My views on the need for this economic landscape stand in some contrast to a fascinating view presented by Haqq-Misra.[125] In considering how to achieve a culturally and politically independent Martian civilisation that, through separation from Earth, can bring forward a transformative influence on human values and outlooks, he presents a series of provisions that would be needed to realise the necessary isolation. To achieve an independence of thought, his recommended provisions involve economic demands including that people on Mars should not acquire wealth on Earth, that technology, resources, or other objects brought from Earth to Mars become permanent on that planet, and that no individuals from Earth may own or claim land on Mars.

The fundamental question that lies at the heart of Haqq-Misra's proposal is whether the economic restrictions he suggests in order to achieve the burgeoning of the experiment in a sociologically isolated and metamorphosed Mars will be more than compensated for by the hoped for creativity and economic efflorescence that will arise in the new society that is born from the conditions so imposed. Although I think that Haqq-Misra's suggestion is an outstanding and bold vision, as the reader may have adduced, I am sceptical about whether such enforced drives to economic autarky in certain activities will yield a sufficiently transformative change to the human condition to make them economically justifiable. Mixing my own doubts about the human capacity to truly change its modes of behaviour, my preference for directing what we historically know about human character to positive ends rather than attempting hitherto untested means of utterly transmogrifying the human condition, and the severe problems that could be brought on by deliberate forms of economic isolation, the trade-off seems to me to be more likely to yield to economic and political tyranny than liberation.

The freedom of terrestrial operators to own Martian land and property and provide for it would naturally inject wealth into Mars. In turn, there would be opportunity for Mars dwellers to own wealth on Earth and multiply and develop it in the context of the vast economic markets to be found on that populous planet, using the wealth so gained to better the economic situation of their home economy on Mars. Such a system seems to me not merely desirable, but a cornerstone of developing a Solar System-wide trading environment.

Economic exchanges of wealth and products may well dilute the autonomy of extraterrestrial settlements to some extent, but at the same time autonomy realised through economic strength can be achieved by the very process of freely trading both people and products with other, and highly economically expansive, branches of human civilisation. Economic interdependency and interaction need not forestall communities becoming places of social experimentation and intellectual novelty. A successful extraterrestrial economy will afford the quality of life and level of provision that can allow individuals and groups to engage in social experimentation, an originality that might be stymied rather than enhanced by the enforced economic isolation of space settlements.

The economic imperatives that emerge through trade both in the extraterrestrial location and in exchanges with other concentrations of human population will drive the formation of a currency system, banking arrangements, and a plethora of other institutional details. In this work, I have given little attention to attempting to predict all the various monetary and financial systems that might emerge in an extraterrestrial settlement; I think that this is about as useful as attempting to prognosticate the financial arrangements of any society on the Earth in centuries from now. I simply want to observe that if we desire the creation of economic systems that are not commandeered under the autocratic direction of single entities, then in general we wish to encourage market mechanisms with all the institutional and financial transactions that are associated with that form of operation.

Although I would encourage free economic activity, we must recognise that corporations can be acquisitive of power and that their capacities to assimilate and coerce are also expanded in their potentialities beyond Earth.[126] Mechanisms are needed to ensure that the commanding corporation or corporations within a settlement never singly control vital commodities including oxygen, water, food, and power and that these four items at least must be subject to anti-monopoly provisions in the constitution or the laws or rules adopted by the society. As with anti-trust laws on Earth, deliberate legal or regulatory provisions, particularly in large settlements where the justice system may allow such provisions, could be implemented. I have suggested mechanisms of 'competitive fragmentation' for the extraterrestrial oxygen industry and other vital supplies.[127] The threshold at which these interventionist behaviours are warranted would have to be determined through experience and prediction. Because they represent interference—essentially central economic planning—these mechanisms should be treated seriously, and with scepticism, and implemented only when a substantial threat to the wider political health and pluralism of the settlement is perceived, or when a monopoly threatens to engulf the only means of production of a vital requirement such as oxygen. A determined and intense focus and study on the conditions for monopoly beyond Earth, and its circumvention without the destruction of economic liberty, is a subject that strongly recommends itself to economic theorists and lawyers.

A feature within the economic sphere that I have not touched upon is the matter of taxation. This is of course a highly contentious subject, and the existence of taxes, their levels, and the way in which they are collected is profoundly important for liberty, not just within a settlement but in its early relationship with Earth. In the case of the American Revolution, it was not merely an important consideration for how the early colonies should be run, but the very motor of the intellectual ferment that ultimately established the basis for separation. Consequent to such history, one might think that this work should address this matter in some detail.

On the matter of internal taxes, in other words those raised within a settlement, I do not think that our knowledge of an extraterrestrial society admits of the possibility of elaborating a detailed taxation scheme, which, much like the monetary and financial system in general, cannot even be predicted for terrestrial populations in the near term. However, later I will provide an argument for the need for a welfare system of some kind, public places for dissent, medical infrastructure, etc. These could all be provided by the organisations running the settlement, obviating the need for any form of taxation. However, we should not discount the possibility that a system of taxation might be used in settlements as a means to engage the population in the public expenditures and to ensure that such monies, in being collected at the point of payment of wages or salaries to individuals, can become a matter of public debate and dissent. Needless to say, taxation can become punitive, or perceived to be punitive, particularly when backed by a governing power's monopoly on the use of force, and so must be subject to the involvement of the population and its agreement.

On the matter of taxation by organisations not located in the settlement, where those taxes are designed for the monetary advancement of people not located in the settlement, one might hope that those who attempt to impose such demands and those subject to them would have learned from history. Ideally the former would realise the likely unpopularity of taxation without a form of representation and the latter would be spurred to action by the considerable wealth of argument repudiating the rights of organisations outside the settlement to impose taxes not used within the settlement.[128] If early organisations are to use taxes as a means to collect monetary gain, they would be wise to ensure that most, if not all, of that harvest is used to advance the interests of the settlement.

History on Earth and speculation on the extraterrestrial case show that taxation need not be punitive or tyrannical provided full voice is given to all those affected by it. For these reasons, I think that the matter of extraterrestrial taxation is an immensely interesting and important subject for future analysis, both in preparation for extraterrestrial settlement and for an indefinite time within the populations that eventually people the spaces beyond Earth.

Redundancy in skills will foster economic freedom

Moving beyond economic transactions themselves, another potential source of economic tyranny is in the manipulation and monopolisation of people, not only the markets they create. In an environment where an individual's essential skills could be a matter of life and death, particularly in the oxygen industry, there could well be a tendency to assign people to tasks and disallow them the capacity to seek other employment because of the essentialness of their task. On Earth, it is not unusual for an individual to be asked to remain in a job while a replacement is found; there is nothing necessarily coercive about such requests. The reality of such situations may be evident early in a settlement's existence, when there are few people to accomplish many tasks. In this case, we might accept the possibility of rules that prevent a person from moving from a position where there is demonstrably no replacement.

What we seek to achieve is a situation where we can realistically build into the political norms and constitution the right to choose one's vocation and to be able to move freely between roles. One approach to this end is to train people with the capacity to do a variety of jobs, allowing them the practical possibility of movement. It is generally likely to be the case that an individual with many skills will be less subject to arbitrary coercion by another person if they have adaptability in the skills they can offer in return for material gain. This policy is further motivated by safety; redundancy in skills prevents the survival of the settlement being dependent on one individual with a specialised skill. However, beyond generating redundancy, I suggest here that training in diverse skills should also be stimulated from the outset by a desire to encourage the conditions for economic liberty.

In a very small outpost with limited options for employment and where the population is confined to a limited number of habitats, movement from one job to another may be nothing more than changing tasks under the same employer. Little expression of true freedom may exist in this situation if the employer exerts a condition of servitude in both jobs. Therefore, this condition also speaks to us of a need to maximise the economic activity of a settlement to generate real prospects for variety in employment.

Even in more advanced settlements, where considerable and complex manufacturing capability has been achieved, the problem of restrictive employment still looms large. The division of labour, a phenomenon central to the capitalistic enterprise,[129] tends to make individuals specialised in small tasks that leave them little able to take their skills elsewhere in the event of unemployment. For this reason, retraining programmes that offer new skills to the workforce to maximise their employability have become a prominent antidote to this problem in modern terrestrial societies. This continuous training in diverse jobs, even for highly trained specialised individuals, might be an educational aim of extraterrestrial settlements.

To summarise, as a general point and principle in the emergence of free government in space, it will be useful for individuals to have capacities in many vocations and skills that give them the maximum potential to offer their abilities to different employers; these options are no less important for the culture of the whole settlement than for the freedom of the individual. If individuals exhibit the capacity to use their diverse skills to change occupation, then organisations are less likely to exert arbitrary coercion on people who are likely to leave them.

The settlement should establish a basic minimal production capacity in vital resources alongside private markets

Although I have placed a high importance on encouraging private markets to ward off excessive economic centralisation, there are aspects of centralised production that I think may be vital for the extraterrestrial case.

Although free markets might well, without any oversight, achieve some minimal level of production in oxygen to ensure that everyone in the settlement has the substance, it would be not merely a market failure but a catastrophic existential threat if it did not. For this reason, it seems to me that one role of a governance structure is not necessarily to coordinate oxygen production, but to be constantly knowledgeable of what the total production capability is and to ensure that it does establish, or keep in operation, oxygen machines that ensure redundancy in production should this be perilously close to the minimum level. This might be less obviously crucial in a location such as Mars, where water and its constituent oxygen atoms, or the atmospheric carbon dioxide and its embedded oxygen atoms, are relatively easy to come by, but in water-poor locations such as the Moon and in the open expanses of space this responsibility conferred on the governance structure can ensure the appropriate level of vigilance. We could view this provision as what we might call 'the core oxygen provision'. We might also apply this liability to water, food, and power production.

As I have suggested, the vitalness of oxygen to human existence could make its provision a fundamental right, and that right could be an essential statement of any constitution, Bill of Rights, or similar document outlining the rules by which a settlement should be run. The governance structures of a settlement would then have the duty to ensure that there exist the means to meet that right for everyone in the settlement. This requirement would add a further level of responsibility to that elaborated in the previous paragraph by motivating minimum oxygen production not merely to prevent a social risk but to realise a right. If the settlement was government-run, then the government would need to ensure sufficient oxygen by whatever means to guarantee a minimum provision for all; if it was a corporate-run settlement, then the corporation would ensure it had the capacity at all times to provide this oxygen; if it was an independent entity, then the settlement authorities, from wherever they originate, would have this responsibility. This oxygen would be free of charge and provided to the occupants of the station by the government(s) or corporation(s) running the station. If the base was run by multiple entities, then those organisations would have to draw up an agreement on how this oxygen would be provided and who would accept the economic cost.

If these requirements were implemented, then we might ask where the free market would have any involvement? Given the vital importance of oxygen, there might be a tendency to draw the conclusion that oxygen must be under total central control. Further, this argument seems to be given force by the physiological realities of oxygen needs. As I have drawn attention to earlier, unlike products subject to fashion and whims, such as clothes, the amount of oxygen required by an individual can be exactly stated given certain assumptions about their physiological state. This physiological exactness of oxygen need might suggest that centrally planned oxygen production, under careful monitoring and regulation, could be both achievable and desirable. However, on this matter, we can perceive a decisive link between the social and political environment and the economic one. Even if we could achieve an exact prediction of oxygen needs and build a machine to produce it, and regulate the operation of this machine to take into account these requirements on a continuous basis, such central economic oversight lends itself to political centrality and the consequent use of this machine or planned network of machines for coercive purposes. The production of oxygen by different groups and organisations lessens, or at least can be used to deliberately reduce, the potential for coercive economic practices and thereby lessens autocratic political behaviour.[130]

I see no reason why, alongside a planned guarantee of a minimal oxygen production capacity, corporations and providers cannot sell additional oxygen machines, spare parts, fashionable oxygen-producing units, and so forth to the population as methods of generating redundancy and plurality in the means of oxygen production. These suppliers could be voluntary possibilities for those who wish to indulge certain fashionable tastes for machines of a certain type, for instance. In essence, such a view is an explicit acceptance of a system of private property. Quite apart from the flexibility that this may allow in the different means of production, a system of private property gives people the capacity to assist other people independent of the designs of a central authority[131] and establishes a domain around a person that is clearly within their own jurisdiction and may be less easily circumvented by others without clear reason.[132] Perforce, this system of economic conduct recognises that humans do have self-interest and are strongly motivated by it, regardless of what utopian ideals we may wish to demand around the oxygen molecule. Adam Smith famously shrunk these matters into his observation that we do not expect the butcher, the brewer, or the baker to provide us with their victuals from benevolence, but rather we address ourselves to their self-interest in making a living in selling these products.[133] Provided that a minimum level of oxygen production is ensured in a settlement, Smith's principle would also apply to an individual who can find a way to make money by selling oxygen machines and their parts, and it should be encouraged. It is precisely the need for a cohesive society under extremity that recommends the animating influence of private markets for the sale and purchase of goods, where private entities compete for the privilege of cooperating with customers to provide what they need for a felicitous life beyond Earth.

However, it is essential that these private means of production do not become an excuse for oxygen production to be left entirely to the marketplace, with the risk of threatening supplies to the population, or certain parts of the population, because of their incapacity to purchase oxygen-producing machines or spare parts. The requirement for oxygen and the calamity that would result from market failure explains why specifying oxygen provision as a right that must be met by those responsible for the settlement could assist in ensuring that we never end up with individuals in the dire situation of worrying about whether they will be deprived of oxygen through non-payment. This latter situation, emerging from an entirely market-based approach to oxygen production, might be defined as the worst, most perverse, and most heinous form of tyranny and human abuse on the space frontier.

The economic environment around oxygen production should never reach a conclusive endpoint of agreement, but it should be a matter of continuous and vigorous discussion, particularly as different locations in space have different potentialities with respect to the sourcing of oxygen atoms, even at local scales (such as water availability in different regions of the Moon and Mars). Specified as a right, it then becomes incumbent on any settlement designers to find the means to realise the minimum oxygen requirement in the location in which they find themselves.

Exactly these same arguments apply to water and food. Although these commodities allow for more flexibility in disruption to their supplies, humans cannot go long without water and food without severe physiological difficulties and social disorder. A settlement's water and food supplies can be open to market development in parallel to some basic planning to ensure that there are entities capable at all times of providing the minimum requirements for existence.

The arguments I apply to these provisions I also apply to healthcare. A settlement, from its outset, will wish to have a central medical facility to handle accidents and other medical situations. The notion of a Mars astronaut having to pay to receive medical attention after an accident with a rover and being left to die on the Martian surface for non-payment is an absurdity. From this trivial thought emerges at once a demand for some centrally provided core medical capability. As a settlement grows, then there is a spirited discussion to be had on whether private capacity could be added to medical infrastructure. Either some type of insurance system or central medical service must be assembled and the preference between these types of provision for a given individual may depend on the permanence of that individual in a particular location, the sorts of jobs they do, and the forms of medical danger associated with particular locations and vocations.

To take all of this together, I simply wish to observe that the extremity of extraterrestrial environments, the criticality of many basic human requirements that emerge from that, and the serious immediate disasters that might be caused by their lack, encourages an economic system where there is a strict responsibility on those who govern to ensure a minimal provision to head off disaster and death. Above and beyond that capacity, a vigorous free market economics should be encouraged and given full play to direct human self-interest to the greater purposes of social felicity, variety, and vigour.

Private land ownership and property is essential for freedom

Such is the effort required to turn part of the Moon, Mars, or an asteroid into a place that ushers forth oxygen, water, and food that it is not clear if the possession of land is a liberating action for an individual. For a settlement to exist, people must transform some land into liveable space by adding a habitat or, in the free plateau of space, building a new space station module. Unlike many places on Earth, the provision of land to an individual does not necessarily set them free. The place of land and resource ownership in the scheme of liberties beyond Earth is therefore a contentious matter. In that an individual must corral impressive resources and collective coordination to turn land into useful living space, we might regard the extremities of the extraterrestrial environment as turning land into a source of potentially illiberal fetters. An individual must join, even subordinate, themselves to a group to have any hope of successfully transmogrifying a piece of real estate into something useful, and this condition applies to a far greater extent than in many places on Earth. Yet, clearly, the possession and transformation of land will liberate a *group* of people in that it provides them with the security to claim some part of space in which to build.

There are those people who claim that land ownership in space is a perversity and that planetary objects and their resources should be the province of all humankind.[134] However, we might remind ourselves of the sheer extremity of the extraterrestrial condition and the difficulty of doing anything productive with land. To discourage in any way the development of these life-threatening extremes by declaring land off bounds for ownership would remove at least one impulse and encouragement to establishing a settlement. This is true not just for expansion beyond Earth, but also for expansion from an existing settlement to another area of settlement on any given planetary body. Those people who develop the initiative to establish a small branch of humanity in a new location should at least be given the liberty of having security of ownership of that land. One could derive this view from an old-fashioned Lockean perspective that having 'mixed' their labour with the land, they have made it theirs: they have transformed that space into their property. However, I do not think one needs recourse to these old arguments to find merit in allowing property ownership of land one has occupied. Instead, the argument might derive from the recognition that the nature of the space environment and the paucity of the most essential products required for human existence that it offers should allow a person or group of people who take it upon themselves to succeed in these forbidding conditions to claim ownership. This view is given force partly because these people deserve to have the security of owning the land they have bothered to work and transform, and partly because this security provides inducement to others to create their own communities.[135]

Although it is the case that resources are not equally distributed in space, it is true that space is a vast realm, and it seems to me a far humbler and less hubristic claim to suggest that people should be allowed to take ownership of a small part of this unpeopled vastness than the counterclaim that the universe belongs to all humanity.

Where people intend to take ownership of rare resources in space that might be contested, an example being ices in lunar polar craters, or where they wish to locate themselves on sought-after real estate, such as permanently lit lunar crater rims, there is obvious merit in those who have the capacity and intention to exploit those resources coming to some agreement on how such resources are to be apportioned.[136] This is a matter for social scientists and lawyers to debate, a process already well underway. In this work, my intention is more to draw attention to the general conditions for liberty in which these claims may exist.

Thus, if conditions for liberty are to be nurtured, it seems to me that private ownership of land must be allowable, either through interpretation of existing laws or through the creation of new international or local regulations to allow for these claims in as harmonious a way as possible. The expansion of private enterprise in this way need not prevent some areas of space, such as certain locations on planetary bodies, from being declared, with broad consensus, as off limits for exploitation: they might be established as parks. These latter areas can be created through motivations that stem from scientific, aesthetic, historical, geoethical, and other rationales.[137]

New challenges in the planning of families and liberty will emerge on the space frontier

I want to now change my focus from strictly political and economic matters to an important social matter, but one that does bear on these former realms. Embedded within the political and civil challenges of new societies beyond Earth is the matter of family structure. In settlements analogous to Antarctic research stations, where there is no permanent population, family planning issues do not arise. However, we can imagine, and indeed we would hope, that if human society is truly to expand permanently beyond Earth, settlements will eventually host families and child-rearing capabilities. In such societies, there is likely to emerge an inescapable tension between the needs of the society and the desires of its members with respect to family planning. Every human being brought into existence will make demands on oxygen, water, food, and power, and of course the variety of other requirements in nurturing and raising children from their early care to their later education. I will have more to say later on the care of children with respect to their education provision and the types of subjects that they might learn.

It is necessarily the case that a settlement cannot have more children, or a population size in general, that exceeds its key production capabilities, and in some circumstances, this may require control on the numbers of children that can be brought into existence. Of course, on Earth few people can simply have the number of children that they physiologically desire, and matters of cost, education, and time all factor into a parent's decision to have children and the number that they might ultimately have; these factors will exist beyond Earth. Yet, in extraterrestrial settlements, there may necessarily be a centrally recognised limit that derives from restrictions in resource availability, particularly in small populations. An objective in the pursuit of liberty is to create the provisions in basic human needs to allow for these restrictions to be minimised to an extent where at least those who wish to have a child may do so and the severity of the restrictions is not so great that this must be denied for most people most of the time.

Since a general objective in any settlement is to create redundancy in the supply of most vital commodities, both from the engineering perspective and from the point of view of reducing the chances of provisions falling prey to single providers, the settlement, if it pursues the tenets of liberty, will likely automatically maximise the chances that minimal restrictions must be placed on the family because of limitations in resources. Nevertheless, we might observe that in planning a settlement and its development, the governing structures should attempt an estimate of the numbers of children that could potentially be brought into existence and thus to plan the expansion of key provisions accordingly. Furthermore, the range of ancillary requirements such as childcare, education provision, and other necessary arrangements must be created and expanded to meet a theoretical demand. The desire to spend time in the effective planning of these potentialities is not only mandated by the need to ensure that the settlement's physical infrastructure is adequate: it is motivated by an interest in producing settlements where the greatest liberty with respect to individuals' family objectives can be realised.

Those objectives in liberty can be partly obtained by the way in which engineering is used to expand the scale and facilities available within a community of settlers.

I have given focus to the freedoms to have children. In contrast to these observations, it could also be the case that an isolated coercive regime could mandate the production of children to reverse a declining population. The right to reproductive autonomy,[138] debarring an authority from implementing forced procreation, has been suggested for space. An obvious limitation of such a right is that if a regime was sufficiently isolated and powerful that it could enforce a population to reproduce, would it respect a theoretical, even constitutionally mandated right? Perhaps greater effort should be placed on building a society where such extreme policies would not be required, or could not be implemented in the first place. Many of the political mechanisms discussed in this work can assuage this concern.

Not unconnected with these matters are future potential developments in reproductive technology. The ability to gestate a human embryo outside of the womb, for example, would not change its oxygen and other vital commodity requirements, but it would provide the opportunity to liberate a mother to be able to lead a life without concerns about carrying a child, worries that may be prominent for people on a dangerous frontier with many high-risk occupations and exploration objectives. We do not yet know what sorts of reproductive technologies will exist on the space frontier, nor do we have any clear understanding of what may be considered ethically desirable or acceptable, yet it is obvious that any such developments will, and must, eventually trigger an energetic discussion on the subject of liberty and childbirth. At the unavoidable intersection of family planning and liberty, it will be necessary for a society to balance the needs for its survival and the demands of a free people.

On the matter of rearing children until their age of independence, I am not a proponent of a kibbutzim-style collective upbringing in which children are removed from their parents to be educated. The strong need for communal bonds to survive in extraterrestrial lethality and the potentially suffocating control of parents or guardians over children in confined spaces might recommend this type of upbringing. However, from Sparta to the Nazis, it has been a characteristic of dictatorial and totalitarian regimes to dissolve or weaken the family unit with respect to state power and to atomise individuals, reconstituting them into a movement or collective. The family unit provides yet another sphere of non-centrally controlled association alongside other voluntary associations that I have discussed. It is a strong and admirable well-spring of thought independent from the state: its existence should be encouraged. Furthermore, except in cases where there is abuse, it provides a loving nurturing focus in a child's life which is threatened with dissolution if education and upbringing are completely torn away from that mooring.

Despite those points, one should also recognise that parents and guardians can themselves exert a claustrophobic influence on a child's upbringing, attempting to propagate ideas and manners from which a child may wish to deviate. For these reasons, it seems to me that early schooling should be conducted in collectives, traditional schools if you will, that offer the opportunity to gain social skills, express dissenting ideas, and build social networks, yet a child should have recourse to a family unit at the end of the day, or on days off. From a social point of view, what we seek is a balance of time split between the care and privacy of a family and the development of a strong sense of collective responsibility to ensure the cohesive success of the extraterrestrial settlement. This balance cannot be quantitatively defined; it must be under constant care and development.

Given the centrality of a cohesive society in the extraterrestrial environment, I think it would be admirable for children to be given experience of that society from an early age. Earlier, I discussed the possibility of assigning citizens to certain tasks, such as participation in independent overview committees, by lot. There may be much to recommend a youth version of this scheme, whereby from a young age, lot would be used to assign youth, for the equivalent of an Earth month a year, for instance, to take part in some assistant function in a randomly assigned part of the settlement. Here they would learn professionalism, the workings of their society, a wide range of useful social skills, and a sense of duty and responsibility to their society. Begun at around the age of 13, the youth sortition programme would give them experience of five different occupations and parts of society by the time they reached the age of 18,[139] helping to prepare them for their adult life and firming up their appreciation of social organisation and its methods.[140] Furthermore, as I shall discuss in more detail later, such experience might help equip them with the necessary virtues to be able to tread the line between liberty and authority, providing the foundations for them to debate that balance with their newly acquired practical experience.

The achievement of independence need not be a defined state of development

I now want to turn my gaze to more expansive matters concerning the liberty of a settlement as a whole and its relationship to other entities. I have said little of the association between space settlements and Earth and how this interaction might impinge on questions of liberty. A matter that has loomed large, particularly in science fiction,[141] is the transition from a settlement established by a founder population to its independence. There is a tendency to project historical terrestrial national independence movements into the space environment and to expect that similar urges will eventually manifest and be made real in space settlements.

It may well come to pass that some settlements in space do achieve independence at some specific point in time with ebullience, fanfare, and noise, but I also suspect that for many, independence may be a long transition period, if it ever occurs in any real way at all.[142] Just as small Scottish islands have their own essentially self-governing structures but still retain strong links to the mainland, the size of space settlements and their isolation may mean that many of them have a mixed form of independence. Governance structures would be run locally to take care of the exigencies of their communities which cannot hope to be run from the distant Earth, yet they would maintain strong economic ties to a planet whose economic scale and populace will always dwarf any space settlement. Indeed, the spatial scales of space that degrade any capacity for a settlement to be run on an uninterrupted basis from a founder nation or corporation situated somewhere else may mean that settlements, from a very early stage, will become politically independent in all but name.

I also think that it is unlikely that settlements might gain different types of independence when they reach certain definite and predictable numerical scales. Intriguingly, US Congressman Newt Gingrich penned a bill in 1981 for space settlements proposing that when they reached a total number of 5,000 inhabitants, they could apply for autonomous governance, and when they had 20,000 people, they could apply for statehood.[143] I think that these transitions will likely be less to do with the numerical size of a settlement and more to do with the capacity for self-sustaining productivity. As the Scottish islands demonstrate, forms of self-governance can be achieved at significantly fewer than 5,000 individuals (on the Isle of Eigg, a mere 100 individuals essentially govern themselves, although the island is not wholly self-sufficient). The desire for 'statehood' will inevitably depend on how the culture of a given size of settlement can establish itself and whether it can do so to an extent sufficient for a population to view itself as something different from any other collection of humans. These are internal matters for a settlement and do not admit of rigorous calculation. On Earth, the Vatican City is recognised as a sovereign state, yet it contains less than 1,000 residents; therefore, even with significantly fewer than 20,000 people, we could expect extraterrestrial societies, once they gain a sense of cultural distinctiveness, to make potentially viable 'states'.

This early sense of independence will be encouraged by the emergence of a culture and set of rules and laws particular to given environments that will ultimately yield a legislative and judicial separation from Earth. At the moment, there is much discussion about the application of existing international and national terrestrial laws to space settlements and their occupants, particularly in the realm of commercial activity; much of this debate orbits the Outer Space Treaty and its provisions.[144] This gravitation towards UN treaties is natural, since there is no permanent settlement from which divergence can occur, so we have only terrestrial legal instruments from which to take a bearing. But once those populations become established, in very short order I see no reason why they should be hidebound to legal strictures established by a Cold War on planet Earth a century or more previously. As soon as they begin to fashion articles of a constitution and develop rules, regulations, and laws to run a society that is aligned to the local conditions, emergencies, and human needs of the moment, they may feel compelled to jettison, or abandon entirely, legal provisions that were devised by Earth-bound lawyers of the twentieth and early twenty-first centuries, and they should feel justified in doing so. The fabrication of this new legal reality will itself perpetuate, out of the origins of any settlement, a form of inevitable social and cultural independence.

Perhaps, from the point of view of liberty, it should be observed that given the vast sums of money, resources, and potentially even human lives that will have been sacrificed in establishing settlements, founder societies or corporations may be unwilling to yield to these various forms of independence if that involves the loss of profits or control. These situations certainly do have their historical analogies on Earth. They are not foreseeable, other than to say that history shows that if, and when, such momentous divergences occur, it would behove the founding group to accept this emerging geopolitical reality and to facilitate such independence to their maximum benefit, rather than to attempt to quash it. Rarely has the latter path been a success. Greater accolades and kudos will be realised by a founder organisation that accepts with magnanimity that it has brought a new branch of civilisation into existence than one that is remembered for exerting all efforts to frustrate its flourishing.

A federated system could nurture liberty at the Solar System scale

In this work, I have mainly scrutinised how governance and its implications for liberty may be expressed within a single settlement. Of course, on the larger scale, we might wonder about governance between settlements and how that might be managed. I suggested earlier that the emergence of tyrannous regimes in space might spread illiberality into other settlements as they take these models as their precedent, leading ultimately to a shadow being cast over terrestrial liberty. There is therefore much to be said for considering how these arrangements between settlements should be managed. Crawford[145] has considered a federated model as one that recommends itself to this task by enabling a plurality of governance structures as each settlement grapples with its particular environmental and resource challenges, yet creating a common bond through which lessons, assistance, and discussion can be shared.

We might think that a challenge that such a structure will face is the effective realisation of a meaningful bond between settlements that are separated across vast spatial scales. These distances mean that the ineluctable limits of physics—that is, the need to communicate using the electromagnetic spectrum—will tend to encourage settlements to operate independently with little capacity to engage in real-time continuous discourse with other settlements or a central federated coordinating body or 'parliament', however that may be convened. However, these physical limits cannot be disposed of (as far as we currently know), and the federal structure appears in principle to be the best system that would be capable of encouraging diverse systems of liberty across the Solar System. The delay in light communication between Earth and the outer reaches of the Kuiper Belt, that disc mass of objects beyond Neptune that might one day be peopled, for example, is several hours, and although this may seem large, it would not obviate decisions between settlements even across these distances over periods of days and months. After all, organisations on Earth function perfectly effectively across international scales in which time zones impose similar temporal delays in communication and decision-making. There is no reason why a federated structure should not gel across the Solar System.

A federated structure lends itself to the variety of idiosyncratic democratic systems that might emerge in different locations in space, these systems potentially providing new sources of cultural innovation for their originating populations.[146] One could also suggest that this type of federated structure, in which settlements enjoy their autonomy but work together to solve their common obstacles, may reduce the chances of conflict, a factor no less important than political innovations. Serious conflict might be ameliorated by strengthening the economic ties and exchanges between settlements, i.e. space commerce, through such a federated structure.[147] The realisation of such a structure would bring into existence a new type of political arrangement, in some sense a genuine cosmocracy—rule over cosmic scales, not merely a parochial pan-terrestrial version of this epithet.[148]

It is significant to this problem that many of the small Scottish islands I have discussed in this work are linked into an organisation, the Scottish Islands Federation (SIF). The SIF has taken on a number of functions. It organises working groups to discuss topical issues common to the islands, such as power production or tourism; it facilitates partnerships and discussions on common projects; and it represents the islands in discussions with the Scottish government. This organisation at once brings to mind the possibility of a similar type of federation to coordinate the interactions of settlements across the Solar System and represent their common interests in discussions with the more populated Earth. The federation is itself part of a wider European Small Islands Federation, which represents over 300,000 people living on 1,640 small islands. This organisation does not see itself as a mechanism of homogenisation, but rather as a means of strengthening the cultural identity of the islands. Its democratic mechanisms, respect for the individual social trajectories of different communities, and capacity to represent many people under a single banner recommends itself as an analogous system for what we could think of as a Solar System Settlement Federation.

The manner in which space at the Solar System level should be explored and settled has been considered by Schmidt and Bohacek.[149] They suggest a 'cosmopolitan' approach, and in examining national aspirations for space exploration, libertarian approaches, and the essentially scientific and technological basis of an extraterrestrial society, they conclude that elements of all these views are needed to make a system of settlements successful. The cosmopolitan view, as the authors explain, is a collection of ideas that advance the notion that humans belong to one community that shares common moral principles; holders of this outlook argue for strong forms of resource redistribution and notions of common heritage that would apply across the Solar System.

I do not contradict the idea that humans should act at some broad level as a community to solve common problems, in the way we must on Earth to solve our environmental challenges. However, the sense of community and demands for the application of common heritage and the equitable distribution of space resources, for instance, must be developed within the framework of the more tyranny-prone extraterrestrial environment. Approaches must be adopted that aid humanity in avoiding autocratic outcomes. Whatever legal and regulatory systems eventually emerge, I have argued in this work that some system of private property is essential from a political philosophical point of view as well as the economic one.

Galactic liberty allows for the identification of universal freedoms

To end this chapter, I wish to make some observations about liberty at its most expansive and extraordinary scale. Turning one's ambitions to places beyond our Solar System, we can see that light-speed delays of years to the nearest stars would make federal discussions quite impractical.[150] In the far-future vision of the colonisation of a local galactic neighbourhood or even a whole galaxy, we could imagine localised federations around specific stars or star clusters. As these federations would be so far apart, not just in terms of electromagnetic communications, but also in human physical movement, there would seem to be no reason for wanting to unite them in any singular federated structure, since such a structure could deliver very little of continuous practical use.

Nevertheless, within individuated federated structures, I see no reason why the considerations discussed in this work would not apply. The rule of law and freedom from arbitrary arrest, for instance, seem to be as eminently applicable on the surface of a habitable exoplanet orbiting some distant star as they are on Earth. The composition and geological evolutionary trajectory of a planetary system is not a factor that moderates whether we allow someone their freedom of expression or conscience.

These thoughts invite and encourage us to consider what the constellation of putatively 'universal' freedoms might be. I would think that freedoms that have no relation to the nature of a planetary system and its location would include: freedom of expression and conscience, the right to assemble and dissent with others about the order one finds oneself in, the right to be treated with respect by others, and when transgressions occur against an established order, the right to be tried (in whatever form that is) fairly and to defend oneself. There are surely others. In that sense, we might well be on firm ground to contemplate, and assert the existence of, truly universal freedoms. These discussions of liberty beyond our Solar System take us to the outer reaches of political philosophy, to considerations on the structure of galactic freedom: perhaps, in some distant time, extragalactic liberty.

4

DISSENT AND WELFARE

Dissent is essential in the extraterrestrial society

In the previous chapter, I suggested some ways in which an open society could be constructed against the tyrannical tendencies that are inherent in the extraterrestrial condition. Yet no matter how well we might organise an extraterrestrial society or how successfully rules and social structures that give voice to people's freedoms might fructify, there will be disagreement.[1] Indeed, a society in which people did not disagree with any forms of organisation would be suspiciously inert. We might say that a society which contains any semblance of freedom is one in which dissent is prevalent, where the various and diverse views of the good life lead to friction and conflict over how those ends are to be achieved. Dissent need not be an infelicity in society as such, but a symptom of a culture that encourages open expression (although it is trivially the case that dissent may reflect deep unhappiness, as dissenters in totalitarian regimes would readily testify[2]). Thomas Jefferson famously said, in expressing a view to James Madison about Shays' Rebellion, 'I hold it that a little rebellion now and then is a good thing, and as necessary in the political world as storms in the physical.'[3]

Given the centrality of the function of dissent in a free society, I want to devote this separate chapter to its consideration, outlining some of the ideas and challenges that its realisation beyond Earth might bring forward and suggesting some ways in which we might successfully ensure its expression.

There is much theoretical thought on dissent and disobedience and the conditions under which it is valid. Here, I will focus on the effects of the extraterrestrial environment, but of course it is still necessary to justify why we think it should be accommodated in an extraterrestrial context at all. Dissent is one means to prevent what Pettit referred to as 'domination' by an authority and particularly arbitrary forms of this domination.[4] If one objective of liberty is to create a space around individuals where they can enjoy some measure of independence and non-interference from the inevitable power structures that emerge in collections of human beings, then a capacity for dissenting to actions that undermine that objective is a crucial feature of a society in which people can protect these very liberties. As Laudani[5] put it:

> Voluntarily accepting a condition of servitude, man gives up his dignity, contradicting his free and rational nature. Disobedience is thus the specific way that men act politically, expressing their natural desire to be free. Understood as a conscious refusal to acknowledge the authority of those who hold political power and to passively serve it . . . disobedience cannot in fact be explained as a direct clash with power but rather as the voluntary withdrawal of support.

I do not deny that this statement is one with which we could argue, but I take it to be the basis of what I discuss here and I will assume that we wish a society to exist wherein these disagreements can be expressed.

In a society which tends towards conformity in extreme environmental conditions, dissent has several functions. First, it acts as a pressure valve to allow people the opportunity to express, or vent, concerns that they may have about the conditions under which they exist. This is a rather simple observation about why open societies allow dissent in the first place. In the extraterrestrial environment, it has the purpose of allowing a pathway for disagreement before it gets out of hand, as Zinn observed for the terrestrial case:[6] 'That is exactly the point of civil disobedience, of a politics of protest—that it is an attempt to bring about revolutionary social changes without the enormous human toll of suicidal violence or total war, which often fall on a society unwilling to go outside accustomed channels.' The application to an extraterrestrial society in a dangerous physical circumstance is at once clear. Second, in an environment where physical extremes might tend to dull expressions of individualism, dissent vivifies the broader culture of debate and discussion. By encouraging divergent views and their realisation, a polity may use dissent to engage the population in invigorated discourse about their society, yielding new solutions, better ideas, and enriched viewpoints about the totality of their situation. It may allow time for the population to reconsider its perspective, even change its view.[7] Third, dissent in the conditions of space can lessen the influence of a central authority and the tendency to centralisation if that dissent is used to bring attention to instances in which that behaviour is deleterious to the population.

Although survival in the limiting conditions of space may have a tendency to discourage and disincentivise a positive view of any forms of dissent, it seems that forms of dissent would be of value to the political health of the society and the self-expression of individuals within it.[8] But how is dissent to be expressed?

People are liable to fear dissent

The instantaneously fatal conditions in any location in space are likely to make people wary of any form of dissent that threatens physical destruction and violence. On Earth, damage to public and private property by acts of civil disobedience sometimes results in loss of life, but in an environment where such damage may either lead to depressurisation or damage to vital infrastructure for oxygen, water, food, and power, violence becomes intolerable. This is the 'problem of violent disobedience'.[9]

The dilemma in dissent arises from the reality that no one can know whether what looks like a peaceful act of disobedience will lead to destructive violence, and this transition, as we know, can be unpredictable and rapid, depending on the temperament of a crowd. The gathering of a group of people who have unhappy intent is likely to cause fear in any settlement.

The authority that runs the settlement will have a strong incentive to quash dissent for fear of being held culpable for allowing a potentially dangerous situation to escalate, but it is equally likely, unless the act of rebellion is widely supported, that the settlement occupants themselves will regard dissent as dangerous. In an environment in which people are precariously poised against an external threat under a welter of technology, unless there is spontaneous and widespread support for action, collective dissent is likely to be viewed with suspicion by all from its inception.

Architecture can mitigate the damaging effects of disobedience

A mechanism by which we might make dissent less dangerous, and therefore less likely to cause a domino-like effect in the social environment in the way in which people disapprove of it, is to make the infrastructure sufficiently secure to be able to absorb the worst possible perceived consequences of violent dissent.[10] This is not to say that we would be encouraging such behaviour and creating a space in which it can occur, but if the engineering structures are sufficiently robust to give people a security that substantial efforts must be made to compromise it, then dissent is less likely to be regarded with apprehension.

The creation of robust physical architecture is commensurate with a desire for redundancy against other acts of destruction such as accidental construction mistakes; thus, in achieving structural redundancy, we also, as a fortuitous outcome, improve the social conditions for dampening fear about dissent. The types of practical implications I have in mind here are habitat walls and materials that can tolerate substantial blunt force without damage and airlocks and other orifices with links to the outside environment being of sufficient security to make depressurisation, even with intentional damage, unlikely. We might also consider the modularisation of habitats that can be sealed off from other areas of settlement, thus localising damage, or ensuring a plurality of production machines for commodities such as oxygen, water, and food so that destruction of some of this capacity does not leave a settlement with a complete want of that vital resource.

The same motives can be applied to deliberate criminal activity, not only unplanned results of group dissent that lead to unpredictable violence. The modularisation and multiplication of oxygen-producing machines, or habitats that are easily disconnected or sealed off from other habitats, would reduce the destructive implications of a single bad action.

Mitigating the fear of the outcome of unpredictable dissent is one aspect we might consider in engineering the settlement;

another is creating spaces in which people can gather to discuss or plan for dissent in private. The existence of widespread systems of video and auditory surveillance in a settlement, which could be introduced to monitor the structural safety of key infrastructure, may become inimical to free expression if those systems are used by an authority to detect and prevent dissent. In this case, rather than introducing a design concept into engineering plans for a settlement, i.e. sufficiently strong infrastructure to resist deliberate tampering and damage, we would be considering removing a potential engineering concept, i.e. camera and audio surveillance, to enhance the conditions for liberty.

I will have more to say on the deliberate attempt to use engineering to expand the boundaries of freedom in a separate chapter, but in the context of dissent we might emphasise the possibility of using engineering to facilitate the social conditions under which dissent can occur.

Release valves must be created for extraterrestrial dissent

Given the interest in preventing dissent from unleashing itself in a single act of destructive rebellion, it would seem prudent for an extraterrestrial society to do everything it can to ensure that those with dissenting voices have a channel through which to be heard. They may not find the power of any given channel satisfactory, but a culture in which a society gives clear and unambiguous opportunities for countervailing views to be voiced is likely to meet with greater approbation by its members. People may be less likely to tolerate destructive actions by a minority when those voices of dissent are perceived to have been given fair play among the wider population.

In the initial stages of a settlement, when a participatory democracy is achievable, the highest forms of power, i.e. the participatory democracy itself, can provide one means for dissent to be realised. As a settlement grows in size and the levels of decision-making may become more representative, these alternative means of independently expressing disagreement become more essential to the group.

However, as with freedom of expression more generally, the larger the number of 'holes' through which pressure may be released, the more likely it will be that voices will feel they are being heard. As Zinn[11] has observed, the objective should be to enrich and expand the capacity for dissent to maximise the non-violent means by which opinions can be propagated into the population, hopefully minimising the perceived need for violence to upend existing social structures in the process.

The plurality of the means of dissent brings us at once back to the wider issues of having the means to express contrarian ideas through media of various kinds and ample opportunity to argue and deliberate points of view; this is as much about the culture of free discussion as it is about the political mechanisms fabricated to allow it.[12]

Beyond these more conventional, everyday means of expression, we could attempt to identify the ways in which dissent might be expressed more forcefully, but without destruction. We might take as a starting point Bedau's[13] definition of civil disobedience, which is 'if and only if he acts illegally, publicly, nonviolently, and conscientiously and with intent to frustrate (one of) the laws, policies, or decisions of his government'. For the extraterrestrial case, the crucial word is 'nonviolently'. Bedau's definition usefully circumscribes the idea of civil disobedience and its objectives. The population of a settlement must have the capacity to act with their conscience. Although breaking the law is a rather troubling criterion—how are any laws to be upheld if no punishment is exacted for anything legitimately claimed as civil disobedience?—the wider point is that society must discern when acts of civil disobedience are intended not as lawbreaking *qua* lawbreaking but as demonstrations of palpable dissent.

Bedau's view of civil disobedience is broadly similar to Gandhi's concept of non-violent resistance,[14] and in the extraterrestrial environment I have suggested that the sorts of action we might be talking about here are approaches such as strike action, sit-ins, and disruption to administrative functions, and possibly even disruption to supplies, provided they do not directly threaten the vital commodities such as oxygen, water, and food in a way that is dangerous.[15]

In extreme settings, a certain respect is required that cannot be easily quantified for the necessities of the community structure needed to survive in those conditions. It is not possible to understand exactly where the line might be drawn between the two states of respect and transgression of respect, but Milligan's set of civility norms[16] might provide guidance for those on the space frontier. They might include general respect for others, rejection of acts driven by hate, avoidance of violence, and avoidance of cruelty. Again, these criteria are themselves potentially nebulous and open to interpretation, but we can see how it is possible to edge towards a society where acts of dissent, even civil disobedience, can be enacted while creating a set of norms and understandings that minimise the chances of catastrophic consequences in confined spaces. Perhaps of special importance in achieving this balance is the provision of education about the history of dissent and its guiding principles, embedded within a broad education on 'liberty', which would equip individuals with the ability to successfully straddle this fine line. I will have more to say about this later.

Extraterrestrial societies must be capable of coping with strike action

Civil disobedience can ultimately lead to strike action, or expressed intentions to strike, in any settlement size. A fundamental question that is raised in the extremities of the space environment is whether this is permissible and, if so, how it is to be achieved without potentially calamitous outcomes. As on Earth, strike disruptions to services and supplies may be unpopular, but not necessarily lethal; interference in an extraterrestrial settlement's supply of a specific and single food item would not necessarily lead to starvation. In these instances, there is nothing on an essential level to be said about strike action other than that it could occur and its end would have to be negotiated.

However, there are realities about the extraterrestrial environment, which we have had many occasions to discuss already, that might make strike action potentially more disastrous. In particular, the extraterrestrial society is riven with many more industries on which human life depends, including the infrastructure and machinery of oxygen, water, food, and energy production, the deliberate shutdown of which, on a large scale, might be perceived to be fatal, even in situations where improvisation might successfully circumvent a want in the short term.

In general, we could predict that in an extraterrestrial environment the occupants of a settlement would want to achieve a rapid resolution to a potential strike or ongoing strike to prevent a deficit of something vital to their existence, and the authorities themselves, charged with the welfare of the enclosed population, will wish to avoid its apparent failure to resolve conflict. Even the strikers themselves, potentially facing the vituperation of a society which they might be placing in danger, will wish to achieve their ends before their actions receive widespread disapproval. There are, therefore, immense pressures to resolve strikes quickly or to prevent them occurring in any vital industry. I have illustrated this previously using the approach of Crouch,[17] and I reproduce this graph in Figure 4.1; needless to say, this is a simplified view to illustrate the way in which the conditions in space could influence perceptions of strike action and thus the liberty to strike associated with it. The strike action, its length, and its proposed means of resolution would depend on many factors, as it does on Earth, including the number of people involved, the temperament and nature of the demands from the strikers, the essentialness of the industry, the prevailing political and economic climate and conditions of a settlement, and the perceived economic effects.

Ultimately, as on Earth, there are few generalised economic points that can be made about the consequences of strike action. Young analysed the consequences of strike action among personnel involved in a putative solar satellite industry in one of few papers to investigate the effects of strike action in an industry located in space.[18] The conclusions he draws about these actions depend on the extent to which the disruption and economic loss would force the operators to meet the strikers' demands, and that would ultimately depend on the cost and profit of beaming energy back to Earth, the cost of solar satellite upkeep, and so forth. Nevertheless, I think the general point—that the extremities of space will likely make strike action more serious in a wider range of industries and potentially more unpopular—is one worth emphasising when we consider the conditions for liberty beyond Earth.

Might there be a case for outlawing strike action altogether given these observations? Apart from the general view that outlawing strike action may be perceived as a repressive measure, potentially encouraging people to take this very route to agitation, it may not be practically necessary to go this far.

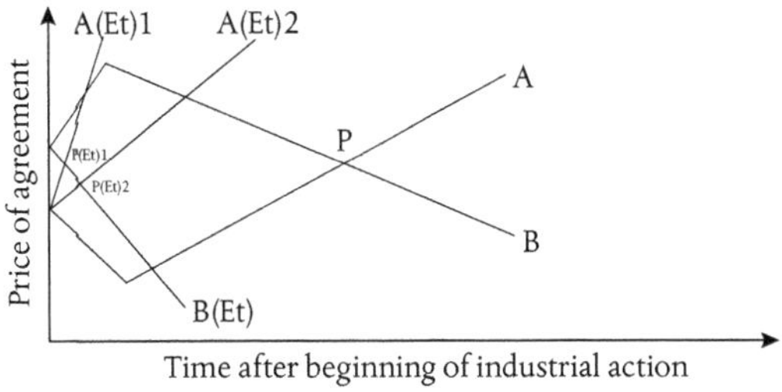

Figure 4.1. Extraterrestrial strike action (after Cockell, 2016b). The graph shows the price (in arbitrary units) that an employer and employees are willing to pay to end a dispute, against time. The curves depicted could apply to a small enterprise of just a few people in a settlement through to much larger corporate activities. For the terrestrial case, line A depicts the cost to the employer over time after the start of a strike. Following an initial period after the strike starts, during which an employer makes a calculation to stand against the strike and their resolve strengthens (the price they are willing to pay to end the dispute reduces), the cost to an employer increases over time as profits are lost owing to the strike action or the implications for failure to return operations to normal become more drastic. The price they are willing to pay to settle the dispute therefore increases over time (line A). By contrast, the employees (shown in line B), initially emboldened by their strike action (and increasing their price for settlement as they dig in), soon find the price they are willing to accept to end the strike reduces as prolonged strike action denies them their livelihood. The employees and employer will reach an agreement where lines A and B intersect (point P) at some time after the beginning of the strike. A hypothetical extraterrestrial case is also shown, say for oxygen production. The price of agreement for the extraterrestrial authorities that oversee the entire settlement will be high (A(Et)1) because they are likely to be held blameworthy for allowing production from a vital industry to be curtailed and will be under great pressure to bring the action to an end. This will vary (e.g. A(Et)2) depending on the product. The price at which employees will settle might be, in some extraterrestrial cases, lower than the terrestrial case (B(Et)) earlier on, particularly for vital industries. This is because the cost to the employees of the shutdown of production of a commodity required in the extreme environment of outer space is likely to be high. It might elicit enormous unpopularity among other people dependent on the products of those industries, increasing the peer pressure to withdraw from such behaviour (of course, the price could be higher than the terrestrial case if those individuals are controlling something vital and have the protracted and wide support from the rest of the settlement). Thus, extraterrestrial strikes in vital industries may be short-lived (P(Et)1 and 2). Of course, these curves are illustrative and would depend on many factors.

Schwartz[19] provides a cogent analysis of the viability of strike action in space. He considers the case of essential services on Earth, such as healthcare, as a model to consider how strike action might occur in vital industries in space. Schwartz makes the observation that industries rarely work in complete isolation and that air production will not be a singular activity but one supported by a network of people involved in this industry, which would include the production of machines, spare parts, and servicing personnel. In other words, the welfare of any given group of people is to a large extent the responsibility of an interconnected larger network. If the overall conduct and organisation of this larger community of people results in some specific group being mistreated or finding themselves working in detrimental working conditions, then it is Schwartz's view that they should be allowed to strike, and it is, in some respects, incumbent on that wider network to create better conditions that would forestall the need to take this action in the first place.

There is merit in Schwartz's solution for the reason that it encourages people within a settlement to take care to ascertain the views and welfare of their fellow settlers. A situation that results in one subgroup of people feeling compelled to strike should encourage others to modify the working conditions. Of course, things are never as plain as we might like. One could flip this argument around to observe that a culture whereby people feel a sense of duty for everyone else could equally give rise to an appalling peer pressure for people to hold back grievances because of the perceived unpopularity they will suffer when other groups associated with them are held responsible for their unhappiness. So if we adopt this wider community view of the acceptability of strike action, then we must also, as with dissent in general, encourage effective communication that allows the aggrieved to bring their views to the attention of others in a credible way, such that strike action is required only in extremis—in a situation where a group feels that all other avenues have been exhausted. This brings us back to the proposed culture of free expression and information exchange which will be so vital to such societies.

If we still think that strike action could lead to potentially crippling consequences for a settlement, then yet another obvious antidote to this possibility is to minimise the chances that a strike would lead to this end. Substitutability in the provision of all vital resources and redundancy in the workforce are means to ensure sufficient resilience in the means of production of commodities. Interestingly, as I suggested in the previous chapter, precisely these motivations apply to achieving redundancy in the production of commodities in case of engineering failures and the more general desire to enhance economic competition and choice in the provision of diverse products. So there are at least four conditions with respect to improving the engineering robustness of a settlement and its general liberties that would suggest a similar goal of enhancing the plurality of production methods: (1) redundancy, (2) economic plurality, (3) resilience to physical disruption caused by dissent, and (4) minimising highly deleterious outcomes of strike action. This is a nexus towards which a settlement should strive, and when we do this, we might find that many of the illiberal, but apparently justified, reasons for controlling employment and banning strike action will fall away.

Revolution will be feared, but can be justified and its effects planned

There may come a time when the behaviour of the governing authority transgresses limits and acceptable boundaries to a large proportion of the population, and so egregious may these transgressions become, and so eroded may the democratic methods of change be, that the population considers that the next step must be revolution.[20] Although less contemplated in the academic literature, this scenario is not new to science-fiction writers; perhaps one of its most famous expressions is Robert Heinlein's *The Moon is a Harsh Mistress,* in which the occupants of a lunar station find themselves embroiled in revolution.[21]

If we were to accept the most pro-revolutionary Lockean view of power—that people have it within their right to remove a government which no longer commands their consent—the question in the extraterrestrial case is how this can be done without the possibility of damage, potentially terminal damage, to infrastructure. Worse still, pre-empting such an end state, an authority may deliberately seek ways to ensure that any physical revolutionary activity will threaten a settlement, thus discouraging actions directed towards the forceful replacement of the authority in the first place; they might place oxygen production in a module or habitat that is vulnerable to potential damage, thus instilling in the population an imminent fear of spontaneous disorderly actions.

There is some difficulty in attempting to establish rules, procedures, and robust statements about revolution in space and its permissibility, because much depends on the way in which a population or group of people attempt such an action. As in the general case on Earth, the legitimacy of such action depends largely on the proportion of the population that supports it. We would likely find the acceptability of a minority attempting a revolutionary coup because they seek power outside the democratic process no more worthy of praise on the Moon than we do on Earth. Perhaps in the case of the extraterrestrial revolution, the special condition that obtains is the extent to which the revolutionary activity results in, or plans to commit, acts of physical destruction that threaten vital survival infrastructure.

A group seeking to implement a change in governance of a revolutionary kind would be wise—a matter that would no doubt be self-evident—to avoid damage that would threaten the existence of the settlement and its population.

Nevertheless, overall we could observe that the extremities of the space environment are not conducive to revolution, particularly violent revolution, compared to Earth. The desire to forestall revolution, or, more exactly, the need for it, recommends all of the mechanisms presented in this work to ensure deliberative governance and the channels to express dissent. If there are means for dissension to be expressed, then we might expect that evolutionary methods of change in society would be more effective and more likely to be realised than the need for revolutionary change. The deliberate inclusion of vibrant routes to dissent will not only allow continuous change and improvement to governance in response to changing ideas: it will also help ensure that society does not become beholden to and trapped in a governance structure that seeks its own preservation to the extent that the only way to change it is by revolution.

The fact remains that ultimately, when a government no longer commands the consent of the governed and the democratic route to its removal has, for whatever reason, been forestalled, the people have the right to remove that government by revolution. I will later address how we might deliberately engineer a settlement to allow for this outcome.

Extraterrestrial societies must have a plan to cope with terrorism

A much more extreme manifestation of dissent is terrorism in space, including malevolent destruction of extraterrestrial infrastructure or human life. These acts do not admit of predictability, and as they are hostile acts, they are potentially directed by an actor external to a settlement. They are not necessarily prevented by good governance within a settlement itself.

For reasons that should now be obvious, terrorist acts have potential for great devastation and loss of life in space. But worse still, the isolation and dependency on many crucial supplies for survival may even make such acts attractive to those who wish to carry them out, since they have a high chance of having a significant effect or drawing substantial attention.

As I have discussed with dissent more generally, one obvious way to mitigate destruction caused by a terrorist act is to design a settlement where acts of physical disturbance have minimal impact. Thus, a settlement does not necessarily have to be designed to stop terrorism as such; a resiliency against such acts may emerge from other engineering and social priorities and principles. Of course, a terrorist action, such as a large explosive detonation or the deliberate flight of a spacecraft into a structure, may be sufficiently violent to overwhelm a structure whose engineering tolerances are designed to absorb a lower-grade cause of damage.

As with terrorism on Earth, a determined actor, no matter what the political and economic efforts might be to prevent such actions, can likely cause substantial damage. Cain[22] has considered in detail the issue of space terrorism, both defining its meaning and providing a variety of examples of the types of terrorist action that might be possible in space. They range from a pathological terrorist act such as the blowing up of a pressurised rover containing passengers to a large act of state terrorism which might involve one state destroying a settlement that belongs to another.

Cain's remedy for many of these potential problems is to develop what is in essence a counterterrorism strategy for space. His checklist of potential mitigation strategies includes: background checks on people leaving Earth (and presumably for people moving between settlements); strict punishment (I address the question of the extraterrestrial judicial system later); licence laws on weapons in space (including biological methods such as genetic engineering and synthetic biology); educational means to encourage rational discourse on dissent (as I have suggested in this chapter, and will do so in a separate chapter on education); and a legislature to implement these types of modalities.

Many of Cain's suggestions align with societies in which education, law, channels of opposition, and other mechanisms seek to allow for, but soften the edges of, dissent. We might therefore argue that the best way to minimise the chances of terrorism in space is to create the sorts of social structures discussed here, i.e. structures that improve the capacity of humans to express their ideas of a good life and to raise protest where those ambitions are perceived to be unjustly thwarted. All the time, we should use engineering to ensure that the infrastructure within which society sits, if it is subject to a violent terrorist act, is physically designed to minimise the damage.

The headache of terrorism is not unrelated to the political structures that form in an extraterrestrial society. If a society solidifies into a rigid monopolistic form of ideology on account of the way it responds to the extraterrestrial environment and this structure becomes unyielding and repressive, then we might imagine that acts of terrorism will seamlessly turn into revolutionary acts to attempt to bring down that structure; a form of domestic terrorism associated with particularly autocratic forms of extraterrestrial governance is imaginable.[23]

We might conclude that there is a broad but important intersection between terrorism and liberty in space. The deprivations that terrorist acts might render in space speak to the creation of maximally open and free societies in space in which designing channels for vigorous defiance is a vital innovation.

A welfare system of some kind will be essential in extraterrestrial environments

Related to the general matter of dissent is the problem of a society containing individuals with an impending lack of something as vital as oxygen. The desperation that a lack of oxygen, water, and food may lead to is not unrelated to the possibility of violent dissent, terrorist-style acts even, but in fact no such extremes are necessary to make this a matter of central concern for an extraterrestrial society.

Not only is a state of penury and destitution undesirable in an extraterrestrial settlement from a purely humanitarian point of view, but the presence of such individuals in a community who lack the capacity for any form of survival outside the confines of the perimeter of a settlement may lead to a range of social ills, ultimately including criminality and then wider forms of dissent against the existing social norms.[24]

In a small-scale settlement, we can expect each individual to be selected for specific tasks, and the small population size essentially militates against anyone becoming destitute. But as a settlement grows in size, there might emerge individuals for whom no task can be found, who are incapacitated or ill, or who for other reasons find themselves outside of the employment and social structures of society. How is this to be managed?

An obvious answer would be to say that one should adopt a policy of full employment. There are, of course, two quandaries with this. First, given the tendency for the extraterrestrial environment to encourage coercive structures of management, we might wonder whether a policy of full employment would degrade into an excuse to control employment entirely, leaving individuals with little choice about their work or the colleagues to whom they are assigned, with all the nefarious cultures of control that this might encourage. Second, even if one could implement such policies in a way that is widely accepted, there would still be individuals who may be incapacitated or ill and could physically do no gainful work. Moving these individuals to Earth may not be practically easy, and an opportunity to do this may not present itself immediately. It seems inescapable that an extraterrestrial settlement must have some form of centrally organised welfare system.[25]

An extraterrestrial welfare system might be organised around the basic precept that individuals should be provided with the means for a comfortable existence when they are not able to offer labour to the settlement. If they are within habitats inhabited by others, it seems that their access to oxygen would be automatic, yet as with the constitutional order, the right to oxygen would ensure that under any circumstances they may find themselves in, no social or economic argument may be used to deprive them of the gas.

A welfare system arrangement might grant a basic income to ensure access to food and other amenities, and regardless of how medical assistance is provided, they should have the means to access it. In other words, irrespective of the specific economic arrangements of a settlement, an individual requiring welfare assistance should be able to obtain liveable standards of oxygen, water, food, and medical assistance.[26]

Of course a welfare system, as on Earth, must not be merely a gesture,[27] but where it seeks to provide for a person, such as with water and food, the entitlement that a person has under such a scheme must be accessible and provide sufficient for that person to be kept in good health. In the terrestrial case, where this is not the case, of course this leads to destitution. In the extraterrestrial case, it will lead to individuals suffering want for vital supplies, potentially leading to antisocial behaviour of serious detriment to society. The exactness with which oxygen, water, food, and power can be calculated, however, should allow for individual needs in case of indigence to be exactly defined and provided. This welfare capacity could therefore be constantly engineered into production capability at all stages of development, and for all sizes of settlement population.

Where a return to work is possible, the settlement authorities should attempt to provide that route back into gainful employment. This is not just a matter of individual health, but, depending on the size of the settlement, a means to avoid the loss of useful skill and work in an environment which, until there is a very large settlement, is likely to find every pair of hands beneficial.

Where permanent incapacity to work occurs, aside from a welfare arrangement, removing these individuals from the extraterrestrial environment into a society that can more easily absorb them, i.e. the populous Earth, may be a preferred option, depending on their ability to adapt to physical and social conditions on Earth.

All of these former considerations also apply to the situation of an ageing population. If we aspire to build permanent settlements in space, then retirement and old age will become a social matter. If we believe that such individuals should have a happy and fulfilled existence, then the capacity to support this non-working population in comfort must be achieved. For a retiree population, the need for support will be more long-lived than for a person temporarily incapacitated. Like long-term disability, this places a demand on infrastructure and requires that long-term support be achievable in the settlement; where that is not physically possible, the transfer of these individuals to larger population centres where such care is available must be planned.

I raise the issue of the welfare system because in the matter of attempting to find ways to channel dissent, the jobless or impecunious must be cared for, as they are one branch of an extraterrestrial settlement through whom a sense of hopelessness may quickly be felt. This sense of alienation may lead to an individual expressing antagonism against society and this may itself potentially contribute to a wider culture of dissatisfaction. Although a retiree population may not represent an imminent source of dissent, they too can give much in the way of knowledge and wisdom to the settlement, and their proper care can do a great deal to encourage a wider sense of satisfaction in the population.

A welfare system or safety net of some kind is called for in any society that takes its responsibilities for the health, happiness, and liberty of its population seriously; but quite apart from that, in the lethality of the space environment, it is part of the edifice of successfully constructed dissent in the extraterrestrial society.

A new type of race prejudice could emerge in space

There are forms of inequality that have been common in societies on Earth that could manifest in space. They are not necessarily linked to the environmental conditions in space as such, but the problems I have drawn attention to before might bring them into existence or exacerbate them. They form part of the landscape of dissent because they potentially bring social rupture into existence.

I will admit to a mildly utopian assumption in this work so far—that people are to be treated with equal dignity and respect in extraterrestrial societies and that all gender, race, and other inequalities are to be avoided. One might hope for this outcome, but here I wish to point out the ways in which prejudices could emerge, and to describe a new form of race inequality that could take root in space.

One prominent possible form of inequality may emerge through race relations. There are the known and historical race conflicts on Earth that could be mirrored in space with respect to the relationships between different terrestrial nation states and regions represented in a population beyond Earth. There is little to be said of this matter other than that one might hope that the professionalism and education of those who venture into space would mitigate these types of prejudices and encourage an emergent cosmopolitan outlook.[28] We work hard to assail these divisions on Earth and there is no reason why the same efforts should not eradicate all forms of terrestrially originated race divisions, and, in parallel, gender and other inequalities, in extraterrestrial societies. For example, provided certain tasks are not assigned to certain genders or terrestrial races with unequal and deliberate intent, the tendency to associate certain types of people with certain tasks, and the prejudices that can result from these perceived differences, can be warded off.

However, there is a new way in which race inequalities might enter into an extraterrestrial settlement that I want to draw attention to. Imagine that in the future there are individuals who have been born on Mars into families that may have reared several generations; they are inured to living and operating in the extreme Martian environment. Newcomers from the Earth will be immature and lacking in the nuanced knowledge and skills to live on Mars; if subtle physical differences exist in an indigenous Martian population, there may even be physical differences of newcomers with respect to carrying out certain tasks in the Martian environment. Perhaps the danger that these new arrivals pose with their limited knowledge will force them into occupations where they represent less of a risk to society. And so from this social arrangement we have the potentiality for a new type of racial prejudice: a prejudice against people from another planetary body. This is racial inequality born of perceived interplanetary differences: it is planetary racial inequality.

As the number of settled locations grows, so the number of possible permutations of racial divisions grows. A lunar settlement has the potential to view Earth dwellers as inferior and vice versa. Settlements on Mars, moons of the Outer Solar System, large city ships and so forth add to the number of two-way possibilities of perceived social differences, leading to a multiplicity of possible racially motivated inferiority or superiority complexes. With just three settlements beyond Earth, there are already twelve possible channels of race prejudice between them and Earth.

I find little to say on this matter because, as all people know, there is no wand to banish these sorts of biases from society. Where true inequalities exist, for example the low adaptation of new Earth arrivals to a hostile Martian frontier, forbearance and gentleness are required, not deprecation and intolerance. As on Earth, the surest way to tackle interplanetary racial inequality is through education, social mechanisms to achieve redress when serious transgressions are made, and an alertness to the emergence of these egregious forms of human inequality. Much the same might be said of gender and other forms of inequality that will not necessarily have novel manifestations in space, unlike racial divisions, but nevertheless could emerge in societies where the cultural conditions have not been alert to their presence.

These same principles apply to disability. It is inevitable that even if the original founders of a settlement were chosen with specific physical capabilities, individuals with disabilities will exist in space, perhaps through birth, but also though occupational accidents. The physical infrastructure of a settlement should be engineered to be capable of accommodating a range of disabilities such as sight, hearing, mobility, and others, either in the extant design, or with the latent ability to be easily modified to allow these disabilities when they emerge.[29] Aside from the engineering considerations, it is naturally the case that we wish to create extraterrestrial societies where human beings are treated with respect and where physical disabilities, as with race and gender, are no basis for the deleterious or prejudicial treatment of other human beings. The way in which we treat our fellow space-faring humans as equals will be an important reflection of the extent to which we have achieved our desired state of liberty beyond Earth.

If we can indeed banish these unjust interplanetary inequalities and eliminate them as far as possible, then a significant moment in human history could well be achieved, since out of these efforts will emerge a true Roman Stoic manifestation of a literal cosmic kinship, wherein individuals become citizens of the cosmos. Humanity will have created the cosmopolis in reality.[30]

Should a free society in space be able to forcefully defend itself?

I want to address a prickly topic, but one which cannot be avoided in a book about building free societies beyond Earth, and even less in a chapter dealing with forms of human dissent. Should a free people in space possess the means to protect themselves by force? This essential question cannot be bypassed, yet it runs up against very strong opinions and thoughts about the prospect of military capabilities in space.

Most often, the debate is framed as an argument about the 'weaponisation of space', a phrase which immediately conjures up a sentiment of extreme opposition, since how could any rational, decent human being be in favour of the weaponisation of space? Let me say that I agree with this general perspective. There are many statements I could make about conflict, war, and weapons that most rational human beings would agree with. Weapons are unpleasant, military aggression is nasty, spending time inventing ways to kill people and using human intellectual effort to do this is a terrible and sad waste of our mental capacities. I think one would be hard-pressed to find humans who disagree with these statements and similar ones.

Yet, these views do not escape the reality that human society gives birth to tyrants and when such individuals and the power that they commandeer threaten to engulf segments of the human population with arbitrary, coercive, and potential violent rule, then the individuals threatened by these social ambitions may need weapons to defend themselves and their way of life. This is a fact that unfortunately butts up against the sentiments I expressed in the previous paragraph, whether we like it or not. It is an unfortunate contradiction between reality and decency that has followed humans throughout their history and arises from the imperfections of people.

Thus, I think it is true to say that none of us approve of the 'weaponisation of space' any more than a sensible person believes in the weaponisation of Earth for the sake of covering the planet with weapons. But this does not solve our problem of how we are to defend ourselves against extraterrestrial tyrants. How is this to be done?

One obvious answer, and one that is common in the space community, is to reply that we must create a strong legal regime in space that has wide international support and can channel and direct national and international space activities to fruitful and peaceful ends. This is a laudable goal, but its contradiction, of course, lies in the fact that tyrants generally do not respect laws. When tyrants do show an interest in laws, it is usually to loosely interpret them or manipulate them to justify coercive behaviour, or to establish new legal regimes that support such behaviour. Indeed, one of the defining characteristics of tyranny is arbitrary coercion or the use of laws to support non-arbitrary but systematic forms of coercive and dominating action, which is often ultimately manifested in a lack of respect for the rule of law or equality before the law. So answering the question 'how do we stop tyrants in space?' with 'by making a good space legal regime' is likely to be self-refuting. We must also remember the facts that we identified earlier: the vast spatial scales of space and the isolation of settlements make it easier for a tyrant to wield power in an environment that, if not locally then at the Solar System scale, may be inherently difficult to police. An agreement about freedom on a pretty piece of parchment in a glass case at the UN in New York is unlikely to interest a tyrant in the Kuiper Belt.

Another answer is to hope that humanity changes its behaviour when it leaves Earth. This is popular among space enthusiasts, and it supposes that the grand vision of space settlement will somehow modify human behaviour sufficiently that once we leave Earth's atmosphere, our past inclinations to disagreement and conflict will miraculously vanish, the power hungry will become conciliatory, and new communal societies of perfect benevolence will emerge. These are praiseworthy and beautiful visions and I do not wish to treat them with undue and destructive cynicism. It is certainly likely to be the case that seeing the Earth from space gives one a perspective on the diminutive stature of human concerns—viewing a small blue planet and appreciating our national borders and historical tribal disagreements as insignificant. Perhaps this is a potentially paradigmatic opportunity to alter the human condition. Even Cicero recognised this potentiality long before there was even a conception of how to leave Earth.[31] In a dream, Scipio is informed by Africanus: 'The spheres of the stars easily exceed the Earth in size. Now the Earth itself seemed so small to me that I felt ashamed of our empire, whose extent was no more than a dot on its surface.' This profound insight—and I am sure the reader would agree that it is remarkable for a mind working in around 50 BCE—has been confirmed by astronauts. It has since been given the title of the 'The Overview Effect', as Frank White's book on this phenomenon was titled.[32] However, there are two sources of doubt that we might have about any optimism that this experience will indefinitely alter humanity. Firstly, we might legitimately wonder whether there is a realistic prospect that seeing Earth from a distance will truly improve human conduct and social behaviours permanently in meaningful ways. Secondly, we might consider the possibility that confronting our insignificance, seeing the Earth so small, could encourage a nihilistic destructiveness in some humans as much as an enlightened generosity in others. We have yet to see what the effects of seeing Earth from space are on a statistically large number of humans to assess these possibilities.

However, to resolve these unknowns, I again turn to James Madison. If space turned humans into angels and banished conflict and the threat of the destruction of free societies forever, then I doubt there is a single human who would not be elated. But we cannot assume that this is going to be the outcome, and a more sensible set of assumptions is that human beings will behave much as they have done so far, that tyrannies will emerge in space, and that those who oppose tyrannical government, either emerging from within the population in which they reside or as an outside threat, must have the capacity to defend themselves.

Does this suggest that societies in space should have the means to defend themselves against tyranny? As a matter of principle, it probably does, and I think that free people in space should not assume that tyrants will acquiesce to their peaceful wishes to be left to their own devices any more than this has worked on Earth. However, fortunately, I also agree that so destructive would the use of weapons and war in space be,[33] particularly given that human societies will be ensconced in pressurised spaces that readily respond to destructive depressurisation, that enormous demands will hopefully emerge in space settlements to minimise the presence of weapons. It would not need conflict to cause disaster; a merely depressed and momentarily suicidal person could wreak immediate and catastrophic damage. Settlements will be aware of that, and presumably social forces and influences will respond in the policy sphere to limit the prevalence of weapons.

One should also draw attention to the potentially ruinous consequences for Earth of malefactors in space. A rock hurled at the planet with a diameter of just a few kilometres would cause a local, and potentially global, scale extinction. Velocities of any object, rock, or artificial object, achievable in space, result in kinetic energies with immensely destructive potential. Humans need to do everything in their power if they are to avoid a single act of civilisation-destroying malevolence, and this includes reducing to the maximum extent possible the militarisation and weaponisation of space. Kinetic energy is the most terrifying instrument of war in space, yet it can also be the nemesis of conflict when its potential enters into rational thought and debate.

Somewhere in here lies a balance between the naivety of proposing that no human being will ever have to defend themselves against tyranny and a dystopian vision of the widespread presence of weapons and the incessant fear that they will conjure in every person living in fragile extraterrestrial habitats surrounded by instantaneously lethal conditions, and no less in those living at the bottom of the Earth's gravity well. As on Earth, I do not think there is some miraculous all-encapsulating policy that can be devised and universally applied; people will need to debate, discuss, and attempt to arrive at a consensus about what is to be done at all stages of human settlement. But in the framework of my concern here—building free societies in the cosmos—I think the onus lies with those who would propose a total ban on all means of defence in space to explain what effective, unfailing, and robust non-violent means is to be used to ensure the complete limitation of autocracy in space and where it does emerge, how it is to be reversed or thwarted in its ambitions to coercively envelop all those around it.

As this is such a contentious topic, let me reiterate to the reader that I am not a hawk proposing the militarisation of space. I deplore conflict and war, but clearly what is needed in our deliberations on space settlement is less utopianism and a sincere and thoroughgoing discussion about how despotisms are to be prevented in space in a way that minimises our need to resort to violent defence. This discussion, the nexus of liberty and its forceful protection, is a very serious and important one for the space frontier, just as it has rightly consumed considerable human intellectual effort on Earth.

Four pillars of freedom: Liberty can be promoted in many of the structures of an extraterrestrial society

I want to complete this chapter with a section to lay the basis for, and introduce, the rest of this work. I have expressed a view that no constitution can ensure freedom. The ideas of liberty, and the desire for it, must take root in the human mind, and to do this we must ensure that the fertiliser and water that can help it grow are in ready supply. These commodities exist in several branches of human activity. No doubt the reader can identify others, and it is true to say that if one thinks sufficiently, one can list a plenitude of social interactions and activities through which freedom can be secured. However, here I want to recognise four key areas of social order that I think are of such importance that I have dedicated a separate chapter to considering each in the extraterrestrial case. These fields of human conduct are the major theatres in which liberty expresses itself on the stage of human history, and through which wholesome dissent is given its liveliest expression.

To begin with, survival in extraterrestrial environments is built on a foundation of enormous scientific and technical capacity. Here I do not claim any categorical or radical departure from the case on Earth, but I want to emphasise that people's lives are utterly dependent on it. If I were to forfeit all of my technological assistance in the British summer and walk naked into a forest carrying nothing with me, I could doubtless exist for a number of days unsupported until I died of thirst or lack of food, or got arrested for public indecency. In the extremities of space, such an action would bring asphyxiation and death, for all intents and purposes, instantly.

In such forbidding conditions, it is essential that the appurtenances of science and technology that sustain life on a second-to-second timescale are amenable to public understanding and that the institutions charged with developing and applying scientific and technological work are constructed in such a way as to minimise the opportunity for authorities to transform the population into a collection of slaves whose lives depend on the technological wizardry of their superiors who work their scientific miracles to keep the people alive.[34] To forestall such an outcome, scientific and technical institutions should be constructed with this idea in mind. In the next chapter, I will have occasion to consider how this can be done.

Rather than simply hoping that science will be applied in ways conducive to free societies, we can recognise that there are engineering solutions that help reduce excessive control. For example, the domination of a single oxygen machine, on which an entire settlement depends, by a king who happens to be a sociopath is likely to be a less desirable situation than a settlement where all individuals have their own oxygen machines, which collectively cannot be controlled by anyone. This *reductio ad absurdum* illustrates that there are engineering solutions that pull society away from the chances of leaving the population at the mercy of coercive authority and create arrangements that are more satisfactory for the happiness of the population. There are many examples of these engineering approaches, and I will return to them in Chapter 6.

Then we have the artistic conduits of society. Its writing, painting, and plays. Its sculptures, metalwork, and theatre. Its poems, songs, and satire. These activities are controlled by the state at one's peril. Particularly in the case of art, human creativity is not easily forced; it can only find its proper and full expression when the human mind is not cajoled and channelled to think and say certain things or, perhaps more to the point, to desist from certain expressions. Yet, there are ways that society can deliberately encourage freedom of expression. By recognising the different roles of the artist in space and identifying the ways in which artistic skills are expressed, we might find purposeful ways of directing those contributions to counter the conformity-causing extremes of the extraterrestrial environment. Such is the importance of artistic forces in the monotonous extremities of environments beyond Earth, I shall consider this challenge in Chapter 7.

We do not know when children will be born in space and so when we will require the first schools, but even in settlements without children, adults will learn and engage in education, just as astronauts today can educate themselves on space stations by reading books or watching videos. Education is a sensitive subject because we run into problems of asking what type of education is objective in any meaningful sense and where that fine line exists between education and propaganda. Acolytes of Karl Marx in the Eastern European nations of the twentieth century no doubt believed that an education in Leninism-Marxism was an objectively good way to understand history and the trajectory of society. Often, it is the benefit of time that allows us to see the prejudice and distortion in ideas that in their time seemed brilliant, novel, and insightful—ripe, in other words, for introduction to the youngest minds. Even against this backdrop of a warning, there may be educational directions and material that find particular use in the extraterrestrial environment—sets of information and capacities, if you will—that may be of benefit to raising children, enhancing in adults a comprehension of their environment and the society in which they live, and furthering the consummation of liberties. In Chapter 8, I will attempt to elaborate on what some of those could be.

So we could conceptually imagine an edifice of liberty built on four pillars: science and technology, engineering, art, and education. These are not exhaustive, but they constitute four load-bearing structures that can prop up the culture of liberty.

At the end of these four chapters, my task is not complete. I will then finally address the justice system and prisons, which have a central role to play in the way society expresses its respect for the fair treatment of individuals who have transgressed in the various and many activities which we are about to discuss and have considered so far. I separate this matter and discuss it last because I consider it to be a capstone to liberty—the system that establishes the legal boundaries of society and handles expressions of freedom that have gone awry. Let us now investigate liberty in four areas of human expression.

5

THE DEVELOPMENT OF SCIENCE AND LIBERTY

Scientific knowledge will be essential for liberty

It is a facile observation that our lives on Earth are underpinned, and facilitated, by the findings of science and their manifestation in engineering. However, when each breath you take depends on a machine, or carefully cultured plant, that produces oxygen, the immediacy of your very existence is completely undergirded by the capacity of science and its application.

There are several implications of the dependency on science in extraterrestrial environments that I want to touch on. The first is the need to understand that science. I will have more to say on education in a later chapter, but here I want to observe the way in which a knowledge of science may be liberating to an extraterrestrial settler.

In commanding the totality of information about the technical edifice that sustains human life, including the details of oxygen synthesis, water production, food growth, and power and electricity production, an authority can take on almost mythical power over the individual standing aloof and awed by this complexity.[1] At the core of the machinery of life is this utter submissiveness to the O_2 molecule, whose artificial production is in the hands of engineers. Now one might say that this is no different from the situation on Earth. Forgetting oxygen, our water and food is produced, cleaned, and processed through networks of technology. To understand water cleanliness, issues of pollution, and the climatic and environmental effects of our food production networks requires knowledge that is bolstered and strengthened by a grasp of science. I think that this parallel is entirely correct and scientific education in our modern highly interdependent and interconnected societies on Earth should be a priority.

In space, the addition of an immediate life-giving requirement into the mix (oxygen gas) and the confinement of all the infrastructure for life within restricted physical boundaries gives scientific know-how and its application a special force. Individuals can only challenge the decisions of those wielding this technology and involve themselves in informed debate with their fellow citizens when they have a basic grasp of science. So the first, most essential, observation we might make is that for an extraterrestrial society to flourish with respect to its liberty, it must educate its population in science and it must encourage open discussion, dissent, and argument about the applications of science and the management of science in the service of survival in that environment.

The second observation is not merely that science saturates the life of a settlement, but that a very great deal of the community's everyday challenges and potential predicaments will be scientific in nature. Be it improvising a spare part for a critical machine, sorting out a structural failure in pressurised rovers, or improving the means to extract water, some of these challenges are technical advancements that occur in the natural course of human social progress, but some of them will be immediate, expedient technical remedies to a crisis or disaster. The discovery and implementation of all of these solutions require the human mind to be able to run free and its impulsive and spontaneous creative capacities to be left to roam unhindered.[2] If this is to be the case, then it follows that the temptation, and a strong temptation it will be, to corral science under central control, and incessantly, and only, direct its products towards specific planned technical objectives, must be avoided. If we accept the role of serendipity in science, mixed with the scientific method and maybe some genius, then this randomness in thought must be given full expression. It follows, therefore, that liberty is inextricably linked to science in that a free mind has more chance of bumping into a previously undiscovered solution to problems and challenges as it strolls across the landscape of scientific possibilities. The people who do this wandering must be trained in science and engineering, which brings me back to my former point about the criticality of science education, but they must also be liberated in the way in which they carry out science. This latter matter brings us to the subject of how we design scientific institutions, which I will discuss shortly.

Scientific activity will be essential if a settlement is to achieve any type of independent self-sustaining existence; every environment in space is different, as I elaborated on in the first chapter. The only way that these environments can be tamed is by the application of human scientific and technical creativity, yet that creative thought must be allowed to channel itself to the specific environment in which it finds itself. A bright new idea for processing perchlorates from soil into oxygen will be met with celebration on Mars and dull disinterest on the Moon. For the collected minds in any location to give that settlement the best chance of survival, they must be free to contemplate the solutions to the physical environment in which they find themselves.

We see in these observations the necessity for science and technology to be directed towards solving local, and sometimes highly specific, challenges, but at the same time not to be hidebound by purely immediate practical considerations. This conflict bears similarities to the timeless discussion on the balance between pure and applied science that many people will be familiar with on Earth, but it takes on a particular significance when the nature of the environment would seem to encourage an unrelenting direction of science towards solving the issues associated with existential threats. Yet despite this enormously powerful inducement to focus on prosaic needs, every effort must be made to construct institutions or bodies of people that can think and act independently, because it is through the unplanned and undirected faculties of the mind that society may find its most effective solutions. To do this, we need to build certain institutions that remain largely independent of the authority that runs an extraterrestrial settlement. I will also explore this matter momentarily, but first I wish to make some points about science and the system of governance.

Science will enhance the means of democratic deliberation

Beyond education, there is a fertile and interesting discussion about the link between science, the emergence of the scientific method, and liberty. The culture that surrounds the application of the scientific method embodies certain ideals, such as arguments constructed independently of authority, objective critique of observations and data, and unimpeded and unconstrained discussion. Although fashioned by the desire to achieve an understanding of the workings of Nature and to probe those mysteries with methods that attempt as best as humanly possible to remove the frailties of bias, these self-same qualities are also ones we recognise in human affairs as those most likely to reduce arbitrary power, encourage open and rational debate, and seek outcomes that are assessed against data and observations, not special interests and political bias. In essence, we might say that the methods of science bear strong resemblance to methods of open deliberative governance.

It is not surprising, then, that much discussion has surrounded the idea that the emergence of liberal democracies in the West has occurred hand-in-hand with the emergence of science since the Enlightenment,[3] and that the establishment of liberal democratic forms of governance has encouraged scientific institutions and the application of science and vice versa. In an inchoate way, this relationship between an open society and the advance of human knowledge has been proposed for the participatory democracy of ancient Athens, and the Renaissance culture of Italian city states prior to the seventeenth-century manifestation of this association in the Enlightenment. I do not think it is necessary to get into long discussions about the veracity of these claims. One might also warily note the profound exaltation of science and engineering by the Nazis as an example of how science was the handmaiden of their quite considerable technical ambitions.[4] Science and totalitarianism have also made a happy couple.

However, it might be the case that a culture that deliberately seeks to encourage open discourse in science and its application may also be one that concomitantly encourages democratic deliberation in the political sphere. In the case of space, Schwartz has made the argument that the scientific enterprise can energise institutions of democracy in space by dint of its need to find solutions to the pressing challenges of surviving in the extraterrestrial environment and the provision and arrangement of services and goods that support open societies.[5] The need to keep technical infrastructure running and the need to improve on the solutions to the production of vital resources require, or at least would benefit from, wide involvement from the population in how to organise, manage, and maintain this infrastructure, quite apart from the detailed scientific and technical operations of its running.

Science does not necessarily give birth to systems of democratic deliberation, but it can facilitate them when it is applied in a system of governance where the scientific effort is a phenomenon with wide public support and involvement. However, crucial to achieving this is the creation of extraterrestrial scientific institutions that are assembled to facilitate public debate and scientific thought.

The extraterrestrial environment can provide the impetus for building new scientific institutions

How can we ensure that scientific work is freely expressed, with the full and enthusiastic engagement of the widest body of people possible? One way is to create scientific institutions that encourage the involvement of the scientifically capable to the maximum extent possible.

On Earth, institutions that accommodate scientific work and propagate it are multifarious and include university departments, government institutes, non-profit foundations, scientific academies, and a wide range of societies. Many of them focus on particular areas of science and engineering, while some represent the totality of science ('national academies'). It would be easy enough to suggest that we should simply create these institutions in space and replicate them there to carry out the same functions that they perform on Earth. However, given the centrality of science and engineering in survival, as I have described, we could also ask how we might create institutions to perform the task of propagating science and scientific understanding, and in particular we might ask how the maximum number of solutions to scientific problems are to be created by involving the maximum number of scientifically capable individuals, while giving them some capacity for free rein to pursue in an open-ended manner the needs of an extraterrestrial settlement.

This challenge brings to the fore a more general question that is applicable to all types of institutions. How should one go about designing institutions in space, scientific or otherwise, and considering their ideal structure? One response could be that we should not design them, but allow their structure to emerge organically through voluntary association. There is merit in taking this view and thereby simply desisting from considering how to design an institution. However, given the conditions that we know exist in space, it may be worthwhile to ask whether there are steps we could take to push institutions in directions that are likely to help them achieve their purposes.

Leaving aside scientific institutions specifically, we might consider how to deliberate on the design of an organisation. There are two ways in which I think one might consider the structure of new institutions in space. The first is to take a terrestrial example and consider how it might be improved upon, how its management structures might be better developed and its general ethos advanced, in the new conditions one is considering. This approach recognises that the longevity of a given institution on Earth suggests (but does not prove) that it has successfully achieved a functional operating format. However, it attempts to improve on any suboptimal attributes that an organisation might possess that could have arisen through error or historical contingency. This procedure has the advantage that it builds on prior experience, but one disadvantage is that an extrapolation of previous structures is likely to carry previous arrangements forward into the new institution without advancement.

The second approach is to start from a blank canvas and conceive of an entirely new structure. This procedure has the advantage that it allows one to achieve some detachment from previous management structures and institutional cultures. The disadvantage is that it might encourage the design of utopian structures whose prior lack of existence might have been because of their impracticality; alternatively, if they have never been tried before for the reason that they were never conceived, they might still turn out to be unrealistic solutions. Nevertheless, this potential dilemma can be overcome by carefully considering any new institutional arrangement and whether it is likely to succeed, and being aware that radically new designs may come with the downside that they might be infeasible.

The advantage of the second approach is that it encourages greater creativity and the likelihood of breaking free of entrenched ideas and assumptions. The risk that structures are impractically utopian is increased, but the chances of advancing entirely novel arrangements that overcome the shortcomings of prior institutions are also improved. It is true to say that we can never be totally detached from considering previous institutional arrangements, since we cannot help but take as some sort of guide what we know on Earth. For that reason, whether we wish it or not, the first approach that we considered above will undeniably find its way in. We will always be in some ways extrapolating or seeking to advance upon what we know already, even if the structure appears entirely new.

We might apply these methods to considering how to create the institutions that will represent science in extraterrestrial locations, whether that be biological societies, physics institutes, or academies representing all of science. Here I will illustrate an attempt to design a science society or academy using the second approach.

We might first decide on the universal facets of such an organisation that we think should exist in any place or time in which it might be formed. Then we consider what would be the ideal arrangements to realise those requirements. We follow that with some due thought about whether this new approach would work or whether we have conceived of an ideal that is utopian but impractical.

For a science academy in the space environment, for instance a lunar or Martian science academy, we could attempt to state some universal characteristics. They might be that: (1) it must represent the totality of science in its membership; (2) it should be effective at galvanising as much of the scientific community and public as possible, both in the environment in which it operates and potentially including people from Earth, given what we have observed about the pervasive nature of science and technology in the matter of brute survival beyond Earth; (3) it should act independently of political institutions for the reasons adumbrated elsewhere; and (4) it should adhere to and propagate the general ethos of the scientific method throughout the population to achieve the wide understanding of science essential to survival in the conditions of space.

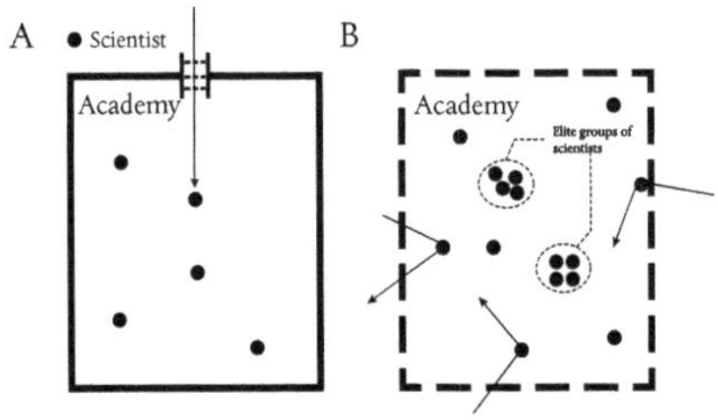

Figure 5.1. A scientific institution in space. On the left (A) is a schematic depicting the typical membership structure of a science academy on Earth. Its means of membership is designed to create exclusivity. On the right (B) is a science academy run on the principle of open membership. The academy is like a town square within which any scientist can wander; here and there exclusive groups of scientists form to debate certain topics in an otherwise open structure. The permeability and accessibility of such a structure is consistent with the wider social objectives within extreme extraterrestrial settings of mitigating the centralisation and opacity of power.

How might we create such an organisation? Existing terrestrial science academies are generally constructed with multiple membership committees that consider, and ultimately elect, members of an academy, thus generating a sharp demarcation between individuals in and outside the institution (Figure 5.1). While this may play to ideas of an institute being an elite body, there is no a priori social principle that requires that a science academy should order itself in this way; indeed, most scientific societies operate on an open membership structure in which some knowledge of the given field is required for membership, but once that knowledge is acquired, an individual can join. There seems to be no reason why such an ethos could not be applied to any scientific organisation.

How would such a principle operate in an extraterrestrial population? The technicalities of such a system are not complex. We could imagine an extraterrestrial academy in which the possession of some scientific level of accomplishment, such as a PhD or professorship, or some defined level of scientific productivity, is sufficient for membership of the academy.[6] That might seem a trite suggestion, but one might ask why this simple concept would be desirable. There are many obvious benefits: (1) the membership criterion is rational and objective, thus any scientist could join and would know what they must achieve to do so; (2) the academy garners the full potential of the scientific community; (3) the academy can avoid any charge of institutional bias, since, as the membership criterion reflects the demographic of individuals with PhDs, professorships, or some other objective criterion, any skew in the demographic in the academy must also be the case in the wider scientific community (this is not to say that a science academy should be concerned about being seen to have bias, but this principle of membership removes any possibility that it can be charged with such bias); and (4) the principle, in its essential nature, conforms to the principles of science: in being objective, it avoids the likelihood of arbitrariness that can accompany selection procedures that are opaque.

There seems no obvious bar to this principle of objective membership. There is nothing impractical or utopian about the proposed structure. One obvious counterpoint is that science academies are supposed to be 'elite', and an academy that admits anyone with a scientific qualification at some defined level is not likely to be elite. If one is to assume that in any given community there are people who are more outstanding than others, which is obviously a reasonable assumption, this principle of objective membership does not prevent elite groups of scientists from gathering within a scientific organisation. A principle of open and objective membership does not obviate the condensation of excellence around specific problems.

What other principles could we develop for a scientific organisation that would approach an ideal? Given the challenge of attempting to palliate the extensive centralisation of power in an extraterrestrial settlement, we could also propose that scientific institutions within a settlement should be encouraged to operate along a principle that we might call the principle of maximum political independence. This principle manifests the more general idea that an organisation should function independently of political influence, that its relationships with other institutions should rigorously reflect the impartial and detached nature of scientific enquiry and the institutional manifestation of this ideal. I even hesitate to call it an 'ideal', since it is readily achievable. The way this can be done is to ensure that the institution has no patronage or formal tie to the governing organisations of the settlement or any other organisations to which it may end up in fealty.

This principle of independence does not negate the requirement to implement the minimum administrative requirements for its existence; this could involve alignment to some rules that the settlement establishes for the operation of the extraterrestrial equivalent of voluntary and charitable organisations. In the early stages of a settlement, such organisations may not exist, and the nascent scientific organisation may find itself in the position of being an independent body, enjoying a largely free and unfettered existence. As the structures of a settlement authority grow and more formal governance instruments come into existence, it may find its financial (and other) activities subject to more oversight, much in the same way as non-profit organisations do in nations on Earth.

A benefit of such a principle is to provide a beachhead against interference by autocratic regimes, should they prevail in the wider political environment—a point that we have seen is of special interest in the extraterrestrial environment.

Interference in scientific fields by autocracies is not a speculative phenomenon. Famously, Elena Ceaușescu (1916–89), wife of Nicolae Ceaușescu, general secretary of Romania's communist party, was a chemical engineer by training, but soon took hold of many prominent positions during communist rule in Romania, including becoming president of the scientific council of the Central Chemical Institute and president of the National Council for Science and Technology; she was responsible for overseeing the country's scientific and technological efforts. The Romanian Academy attempted to push back against this encroachment and was then largely stripped of its powers. Elena Ceaușescu was held responsible by some people for the destruction of much of the fundamental science in Romania in the communist years.

It is not necessary to point out, as Elena Ceaușescu so efficiently demonstrated, that all principles, mechanisms, and cultures embedded in a science organisation (or any institution, for that matter) will always be flimsy against a determined dictatorship. However, the point is rather to incorporate these mechanisms in an institution, not necessarily for the worst-case scenario one can imagine, but against the creeping infractions and temptation to control that the lack of such mechanisms can inadvertently encourage.

Like the principle of objective membership, this principle of maximum political independence is not only a means to achieve the construction of a scientific organisation that properly reflects the culture of science to the wider society, but it may have a practical use as a mechanism to ward off undue interference and as a general bulwark in any period in which governing power is inadvertently or purposefully expanded.

As with other principles, the final step of this consideration must be to deliberate on whether the ideal is achievable. There seems no obvious barrier to its realisation. In the extraterrestrial case, an organisation formed in a largely new society without the presence of existing well-formed and powerful state institutions—a situation that science academies on Earth have not enjoyed—has every chance of starting from a position of maximum political independence and maintaining this position thereafter. However, even if it did not benefit from this extraterrestrial blank slate, the principle itself does not seem unrealistically ideal. Of course, where such an institution is formed in a society in which the governing structure is so coercive and interfering that the organisation cannot realistically achieve this goal, then there is not much to be said on this matter. However, in any reasonably free society, it seems that a scientific organisation can take it upon itself to fulfil its legal requirements while deliberately avoiding all other state affiliations and alliances that compromise its scientific independence. In an enlightened and scientifically educated society, the governing authorities would act with alacrity to help an academy achieve this set of conditions.

In this chapter, I have considered the matter of one of the most cardinal foundations of an extraterrestrial society: its scientific and technical capacity. Given the essentialness of activities in these areas to human survival, they provide a particularly pellucid example of the general means by which we might consider the structure of new institutions beyond Earth and some methods by which we might deliberate and design their means of operation and the principles they might follow. Aside from science and technology, these general principles could be applied to any novel institutions constructed beyond Earth.

Scientific competitions would advance the use of science in the service of liberty

It should cause any scientist some discomfort to engage in a discussion about using science to systematically advance liberty, because of the politically contested nature of liberty and the worry that science will be subverted to purely political ends. Yet there are certain freedoms that can be objectively advanced by science. If I design a pressurised rover that can travel twice the distance on Mars of a previous model using the same quantity of methane fuel, I have enhanced the freedom of movement of its operators by increasing the distance that they could travel on a science or exploratory mission. If I build a spacesuit that uses thinner materials than a previous model and allows more flexibility in the limbs, I have objectively enhanced the capacity for bodily freedom of movement in a real way. In the next chapter, I will discuss further the concept, and potential use, of 'freedom engineering' and its particular application in the extremities of extraterrestrial environments.

Here, I want to suggest that scientific institutions and societies as well as other forms of voluntary association might advance the cause of freedom by promoting competitions and prizes for scientists and engineers to contrive methods, instruments, and equipment that assail the restrictions imposed on human liberty beyond Earth. They might find ways to use the sciences and engineering to extend our capacity for free physical movement in the universe and reduce any other restrictions placed on human physical and mental capacities by extraterrestrial conditions. Where these efforts advance some political point of view or are being used to develop some interpretation of liberty that has no clear influence on physical capacities, we should be apprehensive about the deliberate political commandeering of science, but there are many forms of scientific and technical advances that transparently extend human abilities.

These competitions and awards would be quite valid on Earth, but I will contend that they could be of special use and value to an extraterrestrial settlement. We could imagine scientific societies organising public competitions and displays of devices that enhance freedom which could attract professional effort and amateur work: they would be as much spectacles of entertainment as serious scientific demonstrations. Not only would these efforts improve the conditions for liberty, but they would provide another avenue to bring matters of freedom to the public mind at large and create healthy rivalries between those with technical and scientific inclinations in their contributions to improving life. These competitions would lend themselves as much to adults as they would to youths. Science and engineering prizes could be monetary prizes or simply accolades conferred by organisations or groups of people who engage in this type of work.

Having discussed the relationship between science and liberty and the ideal structure of scientific institutions, I now want to turn more purposefully towards the application of engineering to advance freedom beyond Earth.

6

ENGINEERING LIBERTY

We can engineer freedom into an extraterrestrial settlement

The built environments in which people spend their lives can have a substantial influence on their behaviour and state of mind.[1] This is hardly a controversial point of view, but it focuses attention on the fact that the liberties people perceive themselves to have, and the freedoms that they feel are inherent in their environment, are fashioned by the hand of human technical work. It follows that, in principle, we could direct engineering in a way that would enhance liberty.[2]

The notion of using engineering to 'plan' for liberty should rightfully elicit concern in any person reflecting on freedom. Once technology is widely introduced, it is usually impossible to 'uninvent' it. So if we should apply engineering to some aspect of society in the hope of enhancing liberty, but through unintended consequences we find that we have provided a lever to tyrants, then we will have done little good. From a more political-philosophical perspective, we might also wonder whether a deliberate attempt to plan for liberty is contradictory to our objective to reduce central control and planning. Does this not bring us back to the time-honoured observation that liberty is best secured by voluntary action and not by the well-meaning and high-minded ideals of planners? Should we therefore encourage engineers to desist from meddling in plans for liberty? These questions and their attendant sentiments are well placed, and in broad scope one might say that engineers, and planners of all types for that matter, should avoid an overzealous interest in the planning of freedom. However, the extremities of space might allow us to identify certain types of freedom that are severely disrupted by the natural environment and which could objectively be overcome, or at least mollified, with engineering approaches. In essence, I am going to suggest that we can use planning to reduce the scope of central planning. I hope that this apparent contradiction will be revealed to be less paradoxical once I have explicated what I have in mind.

Throughout this work I have explored many of the factors that act to restrict liberty. We might include, in particular, restrictions on movement, both locally and across the Solar System, and severe technological dependence on the basic commodities for existence. These factors are the inescapable essence of living in space. If we supposed this to be the case, then we could reliably speak of using engineering to lessen their effects or, in other words, to plan for liberty.

I therefore take it to be the case that we can use engineering to build infrastructure that seeks to minimise the effects of the physically imposed restrictions in space and to maximise the capacity that people have to express their lives without these restrictions. Where engineering is used in this way, I have referred to it as 'freedom engineering'.[3] Some people might find this phrase irksome, but I use it strictly to mean the use of engineering to maximise liberties we believe to be objectively useful, and I will later elaborate on how we might pursue such objective measures in engineering. I use 'engineering' here to refer to any form of artificial structure or device.

I have previously defined freedom engineering as: 'an approach to engineering whereby if two technologies are identical in their capacity to deliver a solution to a practical problem, the solution chosen is the one that has the minimum use as an instrument of coercion'.[4] Although I think this does capture the idea of introducing a political-philosophical aspect into engineering beyond only considerations of physics or redundancy, I would be inclined to go beyond this definition to emphasise that in addition to merely comparing two solutions in this way, in some matters we might actively seek out ways to minimise coercion by inventing novel engineering solutions.

The ideas that I discuss are not specific to outer space. I see no reason why one should not think about individual liberties when designing anything on Earth, but I give this field of engineering a name to emphasise how much additional focus we might give to it in environments where instant lethality and its severe restrictions on human freedoms are the norm.

It is worth drawing attention to the fact that there is a vast literature that addresses the psychology of confinement in environments that include Antarctic research stations, submarines, simulated Mars stations, and, of course, past and present space stations in Earth orbit.[5] The use of engineering to alleviate a sense of isolation, and to improve human wellbeing in these situations, would fall under the aegis of freedom engineering, in that we are designing habitation to create beneficial human social and sensory environments. These efforts are certainly a part of the condition for human liberty, but in this chapter I am going to focus on freedom engineering vis-à-vis the removal or mitigation of tyrannical influences, as I have defined it in this work.

I stress that engineering is not a panacea and it cannot entirely remove restrictions caused by the extraterrestrial environment or provide complete liberation from them, but it is yet another potential tool in the arsenal of extraterrestrial freedom, another of the varied and many ways of making the space environment more capacious in its freedoms for the people living in it.

Various types of infrastructure are amenable to freedom engineering

I identified the control of oxygen as one particular challenge in ensuring that the infrastructure for survival does not become used as a means for autocratic control. We can alleviate this problem through engineering (Table 6.1). The situation we might want to avoid is the dependence of a whole settlement on one machine that can be controlled as the single conduit for this essential gas. We might define as one engineering objective the desire to multiply oxygen machines, and even to put them into the hands of the people by allowing ownership of individual machines.

We could then apply these considerations to a specific location such as Mars. There are a variety of ways to produce oxygen on Mars. It could be made by the electrolysis of water originally obtained from the atmosphere or permafrost, by extracting the oxygen from carbon dioxide in the atmosphere, or through photosynthesis, such as from plants, algae, or cyanobacteria. It could be produced by a combination of the above.

Which of these processes offers the fewest avenues for coercive control? The production of oxygen from permafrost depends on the extraction of that material, its melting, filtering (cleaning), and electrolysis, in a system that itself requires oversight and spare parts. We could observe that the extraction and processing of

Table 6.1 Some examples of freedom engineering

Scale of freedom-engineering project	Freedom-engineering item	Restriction being addressed	Objective of freedom-engineering project
Solar System scale	Spacecraft	Lack of freedom of movement in the Solar System and exit from settlements	To make spacecraft reliable, fast, and abundantly available to enhance movement of goods and people across trade routes and enhance exit and entry options
Settlement scale	Habitats	Tendency for central control of vital provisions	To modularise habitats
Settlement scale	Habitats	Fear of damage caused by unpredictable dissent, deliberate criminal action, terrorism	Specifications and tolerances that can cope with physical damage
Settlement scale	Habitats	Sense of confinement	To create large open spaces that enhance sense of freedom
Settlement scale	Habitats	Lack of privacy	To create private spaces that are not monitored

Continued

Settlement scale	Oxygen, water, and food production	Centralisation of production to single items	To create a plurality of production capability: machines owned by individual people, not only by a central authority
Settlement scale	All habitat structures and internal design	Restrictions on disabilities	To enhance operation and movement in settlement for people with diverse disabilities
Settlement scale	All habitat structures and internal design	Intransigent unremovable authority or government	To create robust structures to enable rebellion and revolution
Mobility outside of settlement	Space suits	Lack of freedom of movement outside settlement	To make spacesuits easily repaired, plentiful, easily put on and taken off
Mobility outside of settlement	Pressurised vehicles	Lack of freedom of movement outside settlement	To make easily repaired, reliable, and available pressurised vehicles
Human body	Human body	Engineering for radiation resistance and other extraterrestrial extremes	Allow for greater radiation tolerance, making movement across planetary surfaces less restricted. Comes with considerable ethical debate

permafrost into oxygen is a difficult process for individuals to achieve, making them dependent on many people to breathe. We could similarly point out the large technical edifice necessary to produce oxygen from photosynthetic organisms. This requires growth chambers, nutrients, lighting, heating, etc. Although individuals can own oxygen-producing plants, they in all probability do not personally maintain a sufficient number of plants to provide all the oxygen required for a single person. Thus, both permafrost and photosynthetic production of oxygen require large collective efforts.

In contrast, a machine that extracts carbon dioxide directly from the atmosphere and cracks it into oxygen may be small enough to be used by an individual. Such a machine already exists: the MOXIE (Mars OXygen In-situ resource utilisation Experiment[6]), on the NASA Mars 2020 rover (Perseverance rover), successfully produced oxygen on Mars in 2021. The 'personal oxygen machine' sits directly between the Martian atmosphere and the production of oxygen, reducing the collective effort required and the number of steps between the resource containing the oxygen atoms (atmospheric carbon dioxide) and the individual breathing. From a freedom-engineering point of view, it may be an optimal solution. This solution does not magically dispel despotism; after all, it could be the case that all of these machines depend on one spare part controlled by a venal corporation. But it is likely to be the case that if many people have these machines, then oxygen production is less easily controllable and the failure of a machine could be dealt with by sharing an available spare.

Freedom engineering does not exist in isolation from other engineering objectives. From the point of view of redundancy and avoiding an entire settlement being dependent on a sole oxygen-producing machine that is a single point of failure, we want to multiply oxygen production. If we build a settlement where the oxygen production capacity is achieved through a range of processes and through many mass produced, easily upgraded machines, motivated by the desire for redundancy, then the freedom-engineering objective is achieved as a by-product. However, freedom engineering might give us additional impetus to take certain engineering design directions. If we wish to minimise the number of processes or spare parts through which monopolistic control over the population might be exerted, then we can design machines whose spare parts are easily replaced, even jury-rigged by individuals if needed, and which do not depend on a highly restricted set of manufacturing processes that can only be achieved by specialist companies. From the outset, the machine is designed to be free of the need for involved human and manufacturing intervention in its upkeep and repair.

There is no universal engineering solution to this set of challenges, and the most suitable solution will depend on the specifics of the task. If the zirconia catalysts that the MOXIE instrument depends upon were to be essential for *any* personal oxygen machine and it transpired that in the future these catalysts were controlled by one individual on Mars owning a zirconia catalyst company, then we might reach different conclusions. These engineering decisions are strongly determined by the particular engineering solutions available and the nature of the economic conditions that surround their requirements. However, we could also observe that these very chokepoints, which might be revealed while implementing freedom engineering, could be actively dissolved by ensuring that there existed more than one zirconia catalyst-producing organisation on Mars. In this way, freedom engineering puts us in a frame of mind to deliberately find, and thus to have the potential to reduce, economic and political constrictions in the production of machines vital to human existence.

Freedom engineering is not a strictly quantifiable set of measurements for engineered structures: it is not like specifying the load-bearing capacity of a wall for a given size of house. It is a way of thinking about engineering, a perspective if you like, that consciously considers the restrictions potentially placed on people in the space environment and the ways in which engineering can be used to remove or mitigate those restraints or to actively enhance the capacity of an individual to act in that environment. In particular, freedom engineering is a desire to use engineering to enhance that sphere of private action and thought that individuals may wish for in a space settlement or ship.

Similar approaches to engineering apply to the design and construction of a spacesuit. In this technical challenge, we are confronted by an interest in maximising the freedom of movement of individuals, another one of the defining and limiting conditions of the extraterrestrial environment. Normally, the design of spacesuits is predicated in functional design criteria and the need to have all the amenities necessary to keep people alive within them, such as gas cycling, thermal control, and provision of water.[7] The number of suits and their ease of servicing, although taken into account depending on the astronauts that will require them and the need to consider what servicing might be required in space, are not usually the primary considerations.

If an aim is to maximise the ease with which individuals can leave the restrictions of a habitat and enjoy freedom of movement, then we might want to increase the number of available suits (so that there is always one available if others have broken) and their ease of servicing (to make them easy to repair, even by an individual). These considerations, as with oxygen-producing machines, are also driven by redundancy: planners want to ensure that if crucial outside repairs are needed to a spacecraft or habitat, then a sufficient number of spacesuits are available for this work. But the motivation is given further impetus by thoughts about liberty. If we were to have a station that contained tourists and we wanted to ensure their complete freedom to exit the habitat, even for a brief stroll, we might want to ensure a wide availability of functional suits well beyond the minimum calculated numbers required to meet demands from redundancy.

A freedom-engineered spacesuit is one that is as unidiosyncratic as possible (in other words, not bespoke), constructed with the minimal number of parts possible, with the minimum complexity and chances of malfunction of its individual parts; it is easily repaired, can be easily put on and taken off by any individual, and is mass produced, making it available at any time in large enough numbers to satisfy everyone present in that environment. In this way, the potential for control over the availability of spacesuits by an authority is minimised.

Whole settlements can be freedom engineered

The concept of freedom engineering can be applied not just to items such as oxygen-producing machines and spacesuits, but to elements of a whole settlement. In the same way that we might want to multiply oxygen-producing machines, we might wish to modularise power, water, and food production units in space to reduce the chances of monopolistic control over vital commodities, a motivation at once in alignment with the engineering requirement to increase redundancy in such supplies (Figure 6.1). The achievement of such distributed and multiplied production capability could be met by building a settlement where instead of a single habitat in which key production facilities reside, the settlement is built from a variety of separate habitats, each of which has its own production facilities as far as is practically possible.

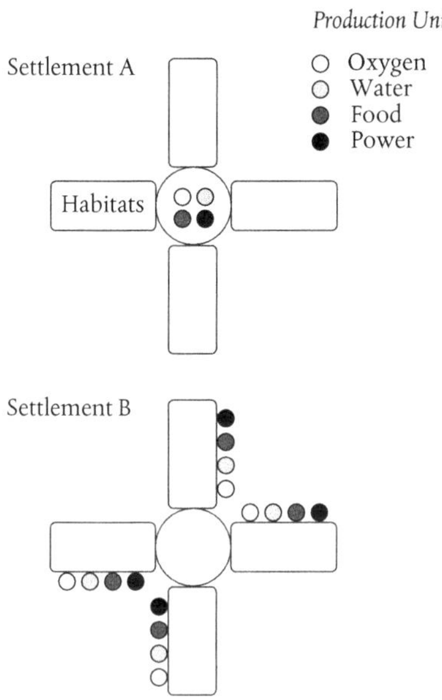

Figure 6.1. Freedom engineering a settlement, for example on the Moon or Mars (adapted from Cockell, 2019). In Settlement A, the individuals are wholly dependent on a single machine for each of their vital requirements—oxygen, water, food, and power—which would leave them susceptible to coercion if those machines were under the control of a single individual or group. However, in Settlement B, there are multiple sources of these supplies, each of which can individually provide the necessary supplies for the whole settlement and each of which are linked into the whole settlement. Although a determined group could control all of these supplies, the task would be more difficult, particularly if those individual units were made and serviced by different groups.

The output of production facilities might be distributed across the whole settlement so that any specific individual's water or oxygen is the sum of a number of production units, making it difficult for the control of any one facility to imply control over a given set of individuals. If each production facility is capable of overproduction, then no failure or control of one facility can threaten the wider population, since other units can increase their output. Again, these precepts are compatible and congruent with the interest in maximising redundancy in a settlement.

Once an infrastructure has been constructed in which the edifice of pipes, wires, and other technical requirements is organised around a unitary production hub, it may be difficult, or at least it will take time, to change the architecture and the culture that have become established around that set of arrangements. The potential difficulty in altering settlement structures at later stages of development emphasises the need to introduce the principles of freedom engineering production facilities at the initial design stage, illustrating how freedom engineering can occur at the earliest phases of settlement implementation.

No social design is without its potentially countervailing effects, and the fragmentation of a settlement for the purposes of devolving technical and production capabilities in the way I have described could equally lead to the formation of factions. The judicious use of central public spaces that connect habitat spaces, thus increasing social activities, but the concomitant deliberate distribution and multiplication of key production facilities in different segments of a settlement are examples of managing trade-offs between these competing needs; in this case, this would involve melding the engineered discouragement of social factions with the deliberate fragmentation of facilities. Quite apart from the desirability of reducing factions, other considerations give force to these conclusions; we would want to have extensive common social spaces in any space habitat to reduce the sense of isolation and separation, with all its attendant psychological and social ailments.[8] As I have already drawn attention to, these public spaces also provide locations for organised dissent.

Finally, an obvious way in which we can freedom engineer a settlement is to expand as much as possible the living space within which people feel free to move.[9] The larger the number and extent of pressurised spaces, the further people can move without coming up against the physical edge of the settlement, and perhaps the larger the variety of biomes and landscapes that can be accommodated. If the extent of a settlement is large enough, it could even become rare for some individuals to be consciously reminded of the reality of the conditions that surround them; the less the reality impinges on a person's mind, perhaps the less they feel that it defines a restricted existence.

The ultimate expression of freedom engineering the environment would be the terraforming of a planet.[10] In this case, the living space would be expanded not by increasing the number of pressurised spaces, but by transforming the atmosphere of the planet itself to be breathable (or breathable with equipment less demanding than a full-body spacesuit). Terraforming is, in this way, a supreme freedom-engineering technology, although its feasibility for given planetary bodies is an open matter. Of course, terraforming would not remove the multitude of questions we have about liberty. Even if we were to engineer a planet to have an atmosphere and surface conditions similar to Earth, all we will have done is to create a stage on which questions of freedom that have stalked humanity since the birth of the first societies on Earth would be repeated under Earth-like conditions elsewhere.

Freedom engineering the right to rebellion and revolution

Stepping aside from introducing freedom into both the acquisition of the basic commodities required for human existence, such as oxygen, and the restriction of movement problem, we could consider how engineering might be used to facilitate some of the more political-philosophical aspects of social order.

I have already drawn attention to the need to create physical spaces for dissent. In some sense, that is a form of freedom engineering. In the extreme case that a governing power has seized control of a settlement and cannot be removed by democratic means, the people within the settlement might resort to the ultimate act of dissent: revolution. If we hold a Lockean contractual view of society, we may even assert that when such a government no longer commands the consent of the governed, the people have a basic right to overthrow it. As the possibility of extreme violent disobedience will generate fear within a settlement, and this fear will itself be used as an excuse for quashing dissent, our objective must surely be to engineer this fear out of the settlement, to engineer the right to revolution.

One approach to this end is to create structures that are strong, can resist blunt force, are robust against depressurisation and rupture, and are modularised such that failure of one module does not imply loss of atmosphere across large areas of the settlement.

The engineering of the structures of the settlement to accommodate violent disobedience should not be done to encourage such behaviour, but we could imagine that a settlement's infrastructure could be designed to allow for this ultimate sanction on power. We could view this as a type of deliberate Lockean revolutionary engineering. Perhaps the knowledge that a settlement has been purposefully engineered to allow for revolution might encourage the acting authorities to seek the consent, and address the pressing concerns, of the population when they know that this lever of political change remains open to those whom they govern. In that way, even if this prerogative is never exercised, the engineering of a settlement to allow for revolution may have a salutary effect on liberty and a tendency to dull the confidence of despots.

Freedom engineering is not always compatible with the need for redundancy

I have made much of the idea that if you engineer a settlement for redundancy, generally increasing the multiplicity and ease of manufacture and repair of various items and production units, you will almost always automatically achieve freedom engineering as a by-product, since redundancy as a matter of fact reduces dependency on, and therefore potential control exerted over, single points of failure. However, there are situations in which we could imagine that functional engineering and redundancy considerations would oppose objectives in freedom engineering, illustrating how it may always be of value to assess freedom engineering as a separate matter from other parts of the engineering process.

Consider the structural monitoring of habitats on the Moon or Mars. The large pressure differential between the inside and outside of a habitat in such locations, coupled with potential radiation exposure and perhaps dust abrasion, depending on how the settlement is constructed, will likely imply that monitoring the structural integrity and the conditions of habitats will need to be an ongoing effort. The way in which this is done is a matter for engineers and may be dependent on the structure, its material properties, etc. But consider a hypothetical example where it turns out that the best way to monitor the habitat is using cameras and audio. Maybe early failure modes in the material are best detected visually as cracks. Perhaps changes in the sound of the Martian wind across the substance is another initial indication of weakness. Now clearly a vast network of cameras and audio equipment throughout habitats can be used exclusively for monitoring purposes, but what if the inhabitants are suspicious, rightly or wrongly, that these monitoring devices are used to track dissent? Perhaps convincing engineering arguments provide a good reason to place these monitoring devices within bedrooms and private eating areas because these spaces happen to be located at the edges of habitats where failures have been observed before.

Here we see an imaginary example of a conflict between people's perceived freedoms and a straightforward engineering requirement to ensure structural safety. If we add freedom engineering into this mix, we begin to ask ourselves whether there might be other ways to monitor the habitats. Cameras are optimal, but perhaps a certain strain sensor would do the job; perhaps electrical conductivity of the material might yield insights on its condition, or maybe we might just reconsider the material choice to avoid a substance for which camera images and audio recordings were the optimal way to track degradation. Thinking about liberty in the context of standard engineering design and planning allows us to identify potential ways in which we might reduce the number of channels through which a more autocratic society could make itself felt. Importantly, here we see a hypothetical example of how freedom engineering yields a different answer from the most expedient solution arrived at by traditional engineering considerations. Thus, liberty is not always an ineluctable product of incorporating redundancy into engineering structures; it should be a separate line of investigation.

Freedom engineering might be applied across the Solar System

Moving beyond the settlement itself to the larger scale of its interactions with Earth and other settlements in the Solar System, we might ask whether similar considerations could apply. One profoundly important factor that leaves humans in a potentially helpless state, and renders an economy potentially detrimentally autarkic, is the isolation of extraterrestrial settlements. Our objective might be to increase the ease with which people can arrive at settlements to bring in new ideas and prevent the stagnation of the population. We might also seek to enhance the ease with which people can leave the settlement. This latter capacity for people to leave settlements to escape political systems and social environments that they find no longer acceptable or bearable has been suggested as the potential basis of a right: the right to leave.[11] In the isolated extremes of space, we can find merit in this idea as a means to subdue an authoritarianism that might become established by encouraging the governing authority to recognise that their poor behaviour has the practical possibility of resulting in exodus. The ease of transfer of goods to and from a settlement can also stimulate the economic condition of a settlement, enhancing trade and preventing a collapse into a highly autarkic economic system in which choices might be very few and economic tyranny of a sort can emerge.

The obvious engineering solution to these economic blights is to create reliable mass-produced spacecraft that can ply the Solar System, exchanging goods and people between settlements. If the capacity exists for regular journeys to and from a settlement, then it becomes more transparent if an authority is deliberately curtailing such opportunities in order to isolate a group from other population centres and restrict the opportunity for individuals to find a way out. Nevertheless, aside from imaginary situations of trying to escape extreme despots, the general point is that even in well-managed societies, the happiness of people and the ferment of economic activity would likely be considerably enhanced by regular and safe services that link a settlement back to Earth and other locations. With these efforts in mobility and free movement across the Solar System, we would witness a general expansion in the liberty of individuals, who would have a variety of options with regard to where they lived and how they traded.

In essence, what I am suggesting is that a Solar System with a larger variety of trade and travel routes is ultimately likely to bring forward a Solar System-wide collection of settlements where people are in general freer in many social, political, and economic respects than in a Solar System of highly isolated groups where people and goods have few, irregular opportunities to move from one place to another. In other words, we can freedom engineer the Solar System and maximise its potential to emerge as a cosmic free-trade zone.

A calculus can provide a means to assess freedom engineering schemes

Engineering is (generally) a quantitative activity, and the word 'freedom' being bolted onto it might cause some understandable discomfort. Isn't this a highly questionable application of a charged and value-laden word with something that has the underpinnings of science? Certainly it is the case that the word freedom is contentious: otherwise this book, and many others besides, would not be necessary. We might wonder, however, whether there are ways in which we could quantitatively assess the benefits that engineering might bring to liberty. Given the values that are inherent in the notion of liberty, I think the answer is, straightforwardly, no; we cannot quantify mathematically and accurately the extent to which engineering can enhance liberty in the same way as we might precisely work out the effect that enlarging a road or a bridge would have on the size of vehicles that could use them.

However, one approach to injecting some degree of quantitative assessment into the use of engineering in this way could be to use a rough calculus that assigns values to certain objectives and then seeks to assess whether, on balance, the engineering procedure is likely to have a detrimental, neutral, or beneficial effect on liberty.[12] Consider the activity of building an oxygen-producing machine that cracks the Martian atmospheric carbon dioxide and procuring these contraptions in large numbers for a population, versus the alternative of a single, central oxygen-producing machine that manufactures the gas by electrolysing large quantities of liquid water.

Let us give positive values to all engineering efforts that enhance the immediate control that the individual settlers have on their source of oxygen (let us say between +1 and +5 depending on the extent to which liberties have been enhanced), and negative scores to all efforts that enhance central control over the oxygen provided to settlers with similar values (our objective being to enhance individual ownership of the means of oxygen production). If we build a central oxygen machine, there may be some enhancement of individual liberty in that individuals no longer have to worry about their own oxygen machines (Table 6.2). Maybe we assign a score of +2. However, there is a substantial enhancement of centralised control over which individuals

Table 6.2 A freedom-engineering calculus

	Oxygen-producing machine that uses Martian atmosphere is individually owned and mass produced	*Oxygen is produced centrally in an electrolytic unit*
Beneficial consequences for liberty	+4	+2
Beneficial consequences for tyranny	−1	−4
TOTAL	**+3**	**−2**

may have little say, so we might assign a value of −4, giving a total of −2. By contrast, individual oxygen machines give substantial control to individuals (let us say a score of +4) but very little central control (let us say −1), giving us a total of +3. On balance, then, we can see that if the maximising of individual liberties and the minimising of central control is our goal, then the engineering of mass-produced oxygen machines is a preferred freedom-engineering solution.

The reader will have rightfully noticed that this is a highly subjective exercise. However, instead of attempting an exact engineering calculation, this freedom-engineering calculus is useful precisely because it does raise basic questions about the assumptions behind it, which can be the subject of further discussion. I have given a score of −1 to the capacity of central authorities to expand their control if we build individual oxygen machines, but what if a vital spare part or material is controlled by a single individual or organisation in spite of the individualisation of the final machines themselves? Such a situation would cause us to question that number. We might instead assign a value of −5; in other words, despite individualisation of machines, there still remains a very strong influence of central control over their manufacture and repair. Perhaps we need to identify all items and production points involved in the fabrication of the machine that are subjected to monopoly and to minimise those effects along the production pathways. The calculus can be a mechanism for discussion and debate that leads to deeper dissection of the implications of processes and sub-processes on freedom, not a dogmatic and accurate statement of the problem.

It is probably safest to observe that freedom engineering cannot be objective. To claim that it is would be to invite a group of people to enforce an engineering solution on a population on the pretext that it will enhance their liberty and that this has been determined empirically and mathematically to be the case, even when the intention could be nefarious. Thus, we might emphatically say that freedom engineering is *not* to be objective, but it is a matter to be debated, while recognising the objective prospect that there are certain engineering solutions that will enhance liberties. An uncomplicated calculus like the one discussed here illustrates that we can develop basic tools and methods that might allow us to contemplate how we can plan engineering to minimise tyranny and start a discourse about those means.

The freedom engineering of a human being

Putting aside the engineering of the human mind, which I will address later, are there bioengineering approaches to modifying humans to give them greater liberty in space? Can we freedom engineer a human? This is obviously a profoundly difficult question ethically, but from a purely technical point of view, would this make any sense? The notion of engineering humans for other worlds has been considered to be an advancement beyond conventional pharmacological treatments for ailments in space because it offers the potential for a more permanent engineering of human biochemical capacities.[13] Here, I want to specifically consider human bioengineering from the point of view of liberty.

We can understand that the multifarious extremes in any space location, notably the lack of breathable air and the potentially high radiation doses in many areas, place a limit on human activity; in short, justifiable anxiety about health and life places a voluntary restraint on human freedom.[14] It does not seem obviously possible that the human body, on its own, could ever be engineered to be able to exist in a vacuum, or even in the relatively benign Martian atmosphere, without a spacesuit. The biological use of oxygen as an electron acceptor in aerobic respiration cannot be replaced by any alternative electron acceptor. Simply put, even if you could replace it with something else, the energy gathered from oxygen-free (anaerobic) metabolism would not supply the human body with the energy it needs to drive its 100-watt power demands.

From the point of view of liberty, the fact remains that humans will likely always rely on a bubble of technology around them to operate in any extraterrestrial environment, a technical carapace that provides the needed oxygen. Perhaps the body might be engineered to cope with short periods of oxygen deprivation or to operate at lower oxygen concentrations than we are used to at sea level, such as by engineering into it a haemoglobin molecule that can bind and deliver oxygen to the body's cells more effectively, as is seen naturally in some people who live at high altitude. But still these modifications do little in extraterrestrial environments, where there is no oxygen to breathe at all. So I am not convinced that any clever bioengineering can release humans from one of the major environmental factors that is the most extreme liberty-restricting condition in extraterrestrial locations: the lack of oxygen.

There may be less significant, but nonetheless possible, modifications that could be made to the human body to relieve some of the restrictions on the space frontier. The engineering of improved radiation resistance into humans, perhaps by enriching our DNA damage-repair capabilities through genome editing, might free us from the need to live underground or within radiation-protected habitats. The reduction in ionising radiation-caused damage might allow people to spend more time on the exposed surface of planetary bodies like the Moon or Mars, or even to visit the radiation-soaked surface of the Jovian moon Europa with less protection, despite the continuing need for spacesuits to provide oxygen. Radiation resistance might allow people to weather the onslaught of solar particle storms without the need to hide in a storm shelter for protection. Engineered radiation tolerance will not cut our shackles, but it could add a modicum of additional freedom to extraterrestrial lives.

Other physiological limits of the human body might be relieved, but it is not clear they would make much difference to the conditions for freedom beyond Earth. Expanding the temperature tolerance of the human body, for instance, will not provide the ability to throw off a spacesuit on the Moon that protects you from temperature extremes well above 120 °C in the lunar day to below minus 170 °C in the lunar night. At the upper end you will boil; at the lower end, you will freeze to death, even if your body was packed with molecules such as engineered anti-freeze proteins.

The unfortunate fact is that many of the conditions of any environment in space are too extreme to be overcome sufficiently by biological engineering to allow any significant extension of freedom of movement and operation. As organic aerobic-respiring creatures, our lives in space are indefeasibly linked to the need for extensive technology to protect our febrile bodies and supply our natural breathing apparatus.

A more radical solution to overcoming these physiological limitations would be to replace biological organs and limbs with mechanical or electronic parts. If one could imagine downloading the human brain and all of its associated cognitive faculties into a machine, creating a human that could escape the need for oxygen gas, then one of the major curtailments in human liberty beyond Earth would essentially be dissipated. Under these circumstances, what we have done is to generate a race of freedom-engineered cyborgs who can now roam freely on the surface of the Moon and Mars in the same way that humans can on Earth. These creatures would achieve new heights in their freedoms to exist in the extremities beyond Earth; they would gain the gift of liberty by electronics and mechanics, freed from the tyranny of biochemistry. However, having done this, they would still have to wonder about their social liberties. As with humans, who have the same capacity on Earth today, this does nothing to remove the basic questions of freedom; it merely has the effect of making questions on extraterrestrial liberty more like their terrestrial counterparts.

Liberty in a post-human Solar System

I have assumed through this work that although the conditions in space are manifestly different from locations on Earth, the essential qualities of humans are largely invariant, and we can expect that those same characteristics will play out in space. However, as a corollary to bioengineering, how would these conclusions change if we could engineer the human brain to behave differently or if we could create a cyborg or 'post-human' entity that transcended our existing nature? Would that not create an entirely different view on the conditions for freedom beyond Earth?

It is, of course, very difficult to make predictions, because we do not know what these post-human developments, if they were to happen, would entail precisely with respect to behaviour, let alone how this would manifest in the context of our familiar concepts of liberty. Indeed, this whole work is an argument about uncertain futures of human liberty beyond Earth, let alone post-human behaviours.

Despite the even greater speculation involved in considering liberty among superhuman entities, let me attempt to make some general comments. As I pointed out at the very beginning of this work, there is no precise balance between collectivism and liberty that can be stated, because society is in a constant state of change, and this flux of ideas, physical infrastructure, and cultures means that the balance between the freedom that individuals have and our desire to maintain collective order of some kind is always a matter of debate and reflection. Even if we were to create a post-human entity or even 'just' engineer the human mind, it is not clear that these new entities could ever escape the need to argue about liberty, because even for them, conditions and society will change.

There are a number of attributes of humans that we could conceivably engineer that might not significantly alter the discussion on liberty as such, but might make it more robust and improve our ability to get the balance right. But what are the knock-on effects on other attributes of the human? Could we engineer away that cloying, grasping need for power and the violence that results when despots do achieve power? Could we create minds that make dictatorship, even if voluntarily chosen by the post-human entities, less destructive? Then we might ask how we could achieve that. If we made humans less aggressive, more altruistic,[15] might this not degrade ancillary desires, driven by the same hormones, to achieve outstanding science or art? We do not yet know enough about the neurobiology of the brain to make these assessments.

We might imagine a mind engineered to have better critical-thinking skills. Maybe when individuals with such minds discuss liberty, they can access and process a vast quantity of literature. Maybe they can read and summarise Adam Smith's *The Wealth of Nations* and Marx's *Das Kapital* in quick two-minute reads. That is all very impressive, and it may allow them to more rapidly come up to speed with literature and simultaneously be able to examine many more factors and ideas that could fashion their view of liberty. They may even be better able to reach a more informed decision. However, they cannot escape the need to decide the balance that Hobbes identified between collectivism and liberty over 300 years ago. They might make a better decision faster than we can, but they will still have to make that decision. But we might still ask whether a vastly expanded computational capacity such as this would necessarily be good? Perhaps this superlative processing power and ability to accumulate longer stretches of the canon of political-philosophical thought make these minds more certain of their conclusions, more demanding of the implementation of their conclusions, more unyielding, and more illiberal?

We might note in a similar vein that a digital superintelligence constructed by humanity and directed towards our problems and conflicts about liberty in space will not necessarily yield liberation from these questions. Instead, the considerable computing power of an artificial intelligence (AI) that can access all of humanity's political philosophy literature might produce solutions that are authoritarian. Perhaps it concludes that eliminating discussions on liberty would itself liberate us from having worries about our freedom. Equally, the far superior power of such a computational device might be used by coercive governing powers to give authority to their plans; liberal forms of governance become a chimera of the frail human mind that the AI can see beyond: its vision of a communistic society in which individual flights of fancy are exterminated is a logical and necessary outcome of the deeper insights and political computations that such a device can make. A tyranny propped up by the calculations of a superhuman computer that ostensibly knows our individual interests better than we do is not far-fetched. Almost every tyranny of the modern age, particularly those advancing central economic planning, has been constructed on the premise that wise bureaucrats can know better what is in the interests of individuals and society than the 'masses' themselves can divine. It takes little effort to imagine a coercive regime shoring up this assimilation of power with a computer that objectively possesses more calculating power than any human brain. Of course, this concern is not specific to the space frontier, but the settlement of space may be contemporaneous with such a potential development, and thus one place in which this possibility could unfold.

Returning to modifying humans themselves, we might wonder about making a brain that is more inclined towards individual liberty, or perhaps the more collectivist among us would argue that a brain that can operate within a collective without a strong need for individualism in the Western sense would be better. If you engineer a human to have proclivities to be a classical liberalist, it is not clear that society will always be better; maybe there occurs a disaster in which one individual with a more accentuated idea of collectivism saves the day. Conversely, if you engineer society to be more collective, it may end up more organised for a while, but perhaps having fewer people with a sense of individuality, and those being less rambunctious in their ideas of freedom of speech, leads to less innovation and more likelihood of a moribund deleterious autocracy. It is not easy to define an engineering goal that would be good for 'liberty' in post-human society precisely because that term is nebulous; and because it is not clear that, even if you did decide on exactly what you meant by that term, the end result would actually achieve what you were expecting.

Perhaps these questions can really only be answered when we have a better neurological comprehension of the human brain and can assess the way in which deliberately altering one facet of human neurological functioning or hormonal regulation would influence other attributes of the human mind.

On either extreme, there lie rocks. A post-human entity that believed in untrammelled individualism might produce a society that collapsed into anarchy; an entity that created an all-encompassing collectivism might produce a society that was efficient but with an extraordinarily conformist dreariness that had very little to recommend it in terms of its creative impulse (although the entities within it might be happy enough). I will desist from claiming that present-day human brains are a faultless and perfect balance—that would be far too convenient—but it does seem to me that, like human society, a post-human society in which there is no agreement about liberty, and the matter is under constant discussion, would be far more flexible and creative and less likely to collapse into these degraded extremes. Perhaps biological evolution has reached that conclusion, and that would be precisely the point.

I am going to take the easy way out of this discussion and declare to the reader that I have little idea what the implications of either deliberately engineering the human mind or producing a new post-human entity would be on the conditions for liberty or how that word is interpreted in such a world, either on Earth or in space. Actually, I will be more defensive and say that this is not so much the easy way out as motivated by the observation that we simply cannot make predictions, such are the unknown links between human physiological and psychological traits and their consequences on the organisation of human societies. If we are so feeble in calculating and predicting the political-philosophical consequences of our own decisions on the future of our societies, how can we hope to make reliable predictions on a currently completely speculative post-human entity and, further, on the space frontier, where we are grappling in the dark to establish how existing human societies will be moulded?

I will conclude this section by simply making the point that it is clearly the case that if the human mind was significantly engineered in some way or derived into a post-human form, the complex interplay of factors that shape the sorts of freedoms that these new societies express will be different in ways we cannot easily discern. But I will complete this chapter with one claim I am willing to make: unless these future intelligent entities are brainless anarchists on the one side or moronic bees connected into an unyielding hive mind on the other, they will argue about liberty. Maybe they will find a use for our books after all.

7

ART AND LIBERTY

The extraterrestrial environment will influence the nature of artistic expression

The human body may be restricted in the scope of its free movement in space, but the mind suffers no such limitations.[1] Of course, it is the case that a confined body may enfeeble the opportunities presented to the mind and curb the breadth of experiences that can bring forth creativity, but in principle the human mind is free to wander where it wishes, even when the body in which it dwells has been shackled by the worst dictator.

Although I have elaborated on the numerous and manifest ways in which people will be physically limited beyond Earth and the ways in which such restrictions may impose on liberty, we might expect that the mental creative capacities would not be so stymied. Instead, one might even venture to suggest that, like great works of literature written in prison,[2] the physical confinement of people beyond Earth may even energise some individuals and lead them to the effusive production of creative art that is given vigour and full-bodied expression by the very fact of their physical restriction.

What sort of creative works can we expect? On Earth, even the most isolated of people have their own styles of art that have emerged over centuries and millennia, changing as circumstances alter and as individuals with particular proclivities set their minds to work. One only has to peruse the types and methods of painting in Europe—primitivism, impressionism, cubism, surrealism, expressionism, modernism, and others—to understand that even in one geographical location, a multitude of forms is possible. This diversity of output is found in equal measure in sculpture, architecture, film, literary fiction, and the other modes of expression. In this chapter, I consider art in its widest sense, covering the many and varied forms of human creative output such as these.

However, we might observe that much artistic expression is given form by the physical surroundings in which people find themselves. Neolithic artists painted large fauna in grasslands; the Inuit of Canada's High Arctic have strains of Arctic life flowing through their work.[3] The phenomenon may be given the obvious explanation that one is naturally inclined to depict the events and features that surround one's home, but it is also perhaps explained by the reality that if one does not travel much, the mind's creativity is always likely to find its most confident expression in things related to one's immediate environment. That does not stop people writing skilful science fiction about worlds they have never visited, but one's local environment is always likely to strongly impinge on what the mind does imagine, and one might observe that even the most bizarre science fiction, behind its façade, takes as its launch point situations that have familiarity on Earth.

With this in mind, we might predict that artistic creativity in space will be strongly channelled by local conditions. Perhaps Martian painting will always have a strong sense of the rocky barren deserts of that world, the profusion of colours expressed largely in browns, reds, or oranges. To the contrary, perhaps those facts will inspire art with a gaudy over-expression of colours not so frequent on that planet—greens and blues—and images and buildings with ever-present depictions of natural ecosystems, lakes, and oceans, features that bring countervailing currents into the lives of the Martians. Similarly, on the Moon, perhaps art will either celebrate the ever-present grey or eschew it entirely. In literature, we might expect a torrent of novelty largely inspired by the desire to escape the physical confines of the space environment; romantic literature or tales of adventure in forests and glens with tinctures of the surreal and unrealistic might be imagined by people who may have never experienced these things or had limited exposure to them.

So much for attempts to prognosticate the nature of artistic expression in space. Yet we can make the prediction that even under the most severe duress from a tyrannical authority, the human mind has never been shuttered and the extraterrestrial environment will give birth to entirely new genres of expression in everything from painting to theatre to sculpture under whatever form of governance is established. Rather than attempt to foresee what art will actually be like beyond Earth, instead we might seek to identify what role artists will play in space and how we might invigorate those capacities for the benefit of liberty more widely.

Art will have many diverse and useful functions in space, including in the propagation of liberty

Beyond the observation that art will be produced in a myriad of forms in space, we might wonder what functions it will perform and whether there are new uses for art in space that have not been so obvious on Earth, or are even irrelevant there. With these applications in mind, we might then ask what could actively be done to encourage and give wholesome expression to such functions in order to ensure an efflorescence of art, but even more importantly, to use art as a mechanism to bring culture and life into the potentially conformity-promoting extremes. Can we use art and artistic expression as a tool of freedom, a means to push back against the physical monotony and restriction-prone reality of living on the space frontier?

In February 2016, while visiting the modern art museum in Stockholm, for no other reason than that I had an afternoon free to do something, I stumbled by accident on an exhibition entitled The Artist. This was not a presentation of works of art produced by famous Swedes, or by anyone in particular. Rather than laying out art according to its age or its creator, which is the normal way to exhibit such material, the museum had taken the unusual step of displaying the pieces according to the perceived role of art in society, with each work used to illustrate a text which sought to explain to the observer a particular way in which art influences, or plays a role in improving, society. This exhibition struck me forcefully. As a non-artist, I had never had occasion to think deeply about the exact roles that art plays in society, other than my uneducated generic view that it was good for creativity and inspiring to look at, as well as providing a useful antidote to barbarism by inculcating something beautiful into society and the people who comprise it. The exhibition had a particular influence on me, because it is hard for me to look at anything new without my mind wandering into space and wondering whether it would have application there. As I walked the corridors exploring the different roles of artists, I took notes about each of them, and after returning to my hotel room, I contemplated their relevance to space, realising that there were categories not exhibited that could be of far more importance to a group of people beyond Earth. That evening, after a brief hotel dinner, I wrote a paper about it.[4]

This way of thinking about art, in terms of its functions, is not just interesting; it provides a means to categorise art and from that position to consider more directly how these uses might be strengthened, encouraged even, to promote the well-being of people in space (Figure 7.1). Let it be pointed out that this idea—to use art to promote a society—is always fraught with worries. As soon as any authority or even small group of people takes it upon itself to direct art, the outcome is often shrivelled and withered, as human creativity cannot be forcefully channelled if it is to be properly called creativity at all. But worse, the idea that human creative capacities should be powerfully commandeered by a state for entirely functional purposes is perverse; it would be to lose the beauty that emerges from the creative expression of the free human mind. Having said that, we should not be dogmatic. Art is always used to express messages, many of them social. Even a prosaic health and safety poster can be a work of art, so there is nothing inherently wrong about art being used functionally. I merely point out that we should not see art primarily as an instrument of state objectives, or else it will be degraded into nothing more than a banality, and the whole purpose of boundary-free human expression will be vitiated.

In the extraterrestrial environment, it should be possible to encourage the uses which I am about to elaborate without demeaning art and turning it into an instrument of power. The recognition and categorisation of artistic purposes need not be an unimaginative exercise in cataloguing, but rather a means to identify the way in which art contributes to the contentment of individuals and the polity as a whole, and then to create the means to give those directions their keenest expression in a

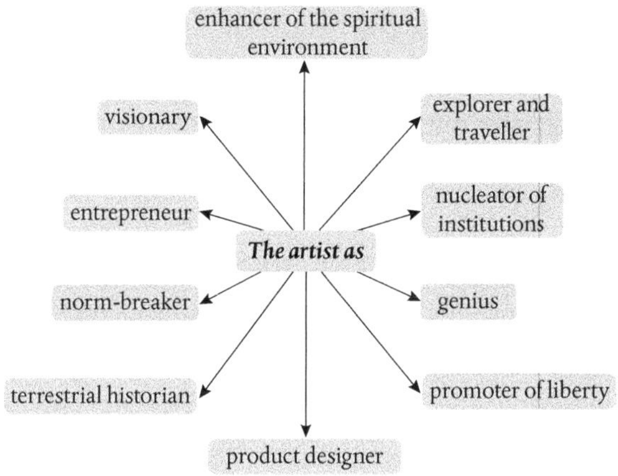

Figure 7.1. The functions of artists beyond Earth. Artists have a variety of functions in creating a free society. Just some of those that can be applied in extraterrestrial environments are shown here.

way that does not interfere with the free expression of artistic adroitness within that particular functional orbit or its particular products.

Perhaps we could begin with a function that would seem immensely important on the space frontier: the artist as an enhancer of the spiritual environment. The term 'spiritual' might seem to some people to be vague, but what I mean by this term is the general feelings of happiness, elation, sadness, melancholy, optimism, pessimism, sublimity, well-being, and all those feelings that shape our emotional lives or might give us a sense of value and being valued, essentially enriching the fabric of our existence.[5] I am using the word spiritual to envelope the totality of our emotional lives and emotional interactions with our environment and other human beings, not necessarily to connote religious ideas, although for some people that may be a part of this fabric. Art has always fulfilled this function in society. Indeed, the reason why many of us go to art galleries in the first place is to engage with our emotional existence, to revel in the unexpected, the sublime, the beautiful, and the abstraction that artistic effort projects.

Beyond Earth, the extremities of the environment and the restrictions they impose bring special force to the purpose of art as a means to lift people's spirits. On the Moon and Mars, for instance, surrounded by grey or essentially red radiation-draped wastelands, respectively, the interior of habitats must take people away from these external realities and give life to those emotions that generally constitute the human spirit. The artist has a consequential role to play in realising this need.

The types of art that we could imagine contributing to this goal are paintings and drawings that raise the spirit, perhaps bringing colours that are rare outside (reds, oranges, greens, and blues on the Moon; blues and greens on Mars) into people's lives. Sculptures and buildings may be fun, humorous, or capture a spirit of soaring optimism. In the literary realm, we might imagine books and plays that make light of the extremities of space, Greek tragedy-like tracts that bring hilarity, perhaps infused with satire and barbed comments about the authorities that run the settlement. Music, too, as on Earth, has much to recommend it: from the limitless horizons of classical music, to the shallow but uplifting lyrics of a popular song, all of this has a role to play in reaching the many and diverse tastes of the human population, but expressing the common effect of lifting people from everyday normalcy. Beyond Earth, songs will become facilitators in liberating the human spirit.

Equally, music might directly confront the human predicament beyond Earth, touching on sadness, isolation, the black eternity of space, maybe the unrequited love of one sent to Mars, separated from a sweetheart on Earth. The melancholy and sadness of the human condition and its affirmation in art is not just inescapable, but it is a catharsis for this ever-present reality; in space, art may well act not only as a distraction, but as a pressure-release valve for these emotions.

On Earth, because of the essentially benign environments in which we live and the life-giving atmosphere, our architecture is diverse in form and shape. It is not limitless; it must conform to the laws of physics, to the rules of compression and tension in the Earth's gravity field. It must tolerate rain, snow, sand, and a variety of other physical factors that manifest in different places on Earth. Yet, outside these requirements, it can find an open field. In space, the environment forces upon designers a functional frame of mind. Spacecraft take on the forms of tin cans, and even their interiors tend to become displays of technological function, wires, pipes, and screens that give their occupants an immediate and continuous view of their essential life support systems. We might notice that, in general, greater environmental extremes impose greater architectural limits and have the parallel effect, given the magnified chances of death, of focusing the architect's mind on ensuring reliable survival, with all the restrictions on design that that entails. This reality can be observed on this planet, specifically in Antarctica. Without the need for a large pressure differential between the inside and out, architectural expression has wider possibilities on that continent compared to space—and some Antarctic stations are stunning—yet their designs are far more circumscribed than the range of buildings to be found in most temperate town and cities.

These limits on architecture in space cannot be escaped; they are mandated by physics.[6] However, they do not necessarily prevent the free expression of the human mind. Even an unimaginative ellipsoid living module on another planetary body could be covered in the façade of a Tudor house; why not the beamed roof and walls of a Tudor mansion adorning a lunar habitat? Even inside, artists can cover the cold functionality of steel and aluminium pipes with walls that bubble with joy and the playful images of forests or the walls of an eighteenth-century palace. The artist would have a much more prominent role in the extraterrestrial environment, working inside and out to bring spiritual grace, humour, inspiration, awe, and a myriad of other human emotions to the technological means of existence. This could be a liberating force in the human future beyond Earth. Often space agencies and organisations eschew these ideas—try suggesting a Tudor country house as the central design criterion to a NASA lunar habitat task group. Suggest to the Chinese space agency that their next lunar lander should be given an exterior cover to make it look like a roaming panda, one of that country's national wildlife symbols.

When national prestige, human life even, is at risk, these ideas seem frivolous, even plain silly. But as humans advance beyond Earth and the question of liberty comes more directly into view, we must let go of these engineering prejudices and mobilise artists towards the objective of alleviating the banality of extraterrestrial extremes with creativity that allows the human spirit to blossom. Yes, Tudor houses and robotic pandas on the Moon.

The overlap of art and human spirituality finds expression in remembering those who have given their lives. At the Apollo 15 landing site, at Hadley Rille on the Moon, lies a tiny 8.5-cm-long aluminium carving of an astronaut. Alongside it rests a plaque commemorating the 14 astronauts who had lost their lives up to the point of its installation in 1971. *Fallen Astronaut* is a work of art, and it was created by Paul Van Hoeydonck, an artist whose work explores the link between humans and technology. Here we see art in the function of the human spiritual need to mourn those we have lost and, as a society, those who gave their lives to achieve a greater purpose for us all. Astronauts have died in tragic accidents since then, and as humans permanently settle beyond

Earth, we can be sure that others will sadly continue to be lost. We can never build perfectly safe rockets and habitats. As space becomes occupied by people other than the highly trained astronauts of yesteryear, ordinary citizens will likely lose their lives too. Artists are ready to remember them.

We might recognise that this artistic spiritual advance is not only for passengers heading beyond Earth. Some of this alien art will find its way back to the human population on Earth who for some considerable time will represent the majority of the human race. Apollo astronaut Alan Bean's paintings were inspired by his trip to the Moon, some of them even being sprinkled with lunar dust from his spacesuit to give them a little added lunar chutzpah (and value too). In this work, we have already witnessed art inspired by another world bringing spiritual enlightenment to the people of Earth. This is not a speculative future function of art and artists; we can already feel and see its influence. Alan Bean's paintings are likely the tip of an artistic iceberg when we consider the paintings, sculptures, literary works, fine buildings, poems, plays, and films that could eventually pour forth from innumerable settlements across the cosmos. Perhaps space artists will re-energise artists on Earth. Not to be outdone by the Johnny-come-latelies on Mars, a gentle but energetic rivalry might emerge between free-thinking minds across the cosmos and on Earth to exceed and outdo previous epochs of artistic endeavour. The whole Solar System as a storehouse and begetter of superlative art from the lone consciousness that inhabits it is a remarkable vision of our human future.

Artists cannot just enliven us in the present; their creative impulses can take us into future worlds, and in this function we find the artist acting as visionary, revealing to us in words and images our potential futures, for good or bad. This is not a purpose that requires humans to occupy space. A search on the internet for 'Moon base', 'Mars station', or 'Asteroid mining' will fill your screen with a never-ending bricolage of imagined human futures beyond Earth, a variety of homes, robots, and expeditions fanning out across the Solar System, and even robotic emissaries and world ships spanning out across the galaxy. This genre of space art is fully developed, and it did not require the launch of humans into space to inspire it; artists were already envisioning the human exploration of space before satellites were launched, including most notably, but not exclusively, Chesley Bonestell's visions of stations on the Moon, Mars, and in Earth's orbit. Remarkable is some of the accuracy of the landscapes depicted by these artists long before the objects of their interest had been visited by space probes and imaged from close up.

Nevertheless, it must surely be the case that once humans have settled other locations in space, artists will get to work envisioning futures that are more in tune with real possibilities in the short term. Perhaps artists will paint renderings of polar expeditions on Mars, new proposed buildings on Callisto, and other projects that are not mere visions for the sake of it, but commissioned and self-motivated works to illustrate emerging plans. The artist can always be one step ahead of the engineers and the population in bringing to life ideas and concepts that are otherwise merely passed in words from one imagining mind to another. They liberate the human mind by giving full expression to ideas that otherwise linger in the skull of a single human being.

This artistic process can be a form of visionary liberation. Artists might imagine new and larger buildings that give wider scope for movement. They might imagine the assembly of settlements on new worlds and in distant locations in space. Just as these works of art in the 1950s represented a sort of escapism at the birth of the space age, a depiction of a soaring vision of the future, so too these works created on future Earth or in other locations in space represent the expression of the free creative urges of humans to project their minds far into the future and envisage what might be and where our species might eventually roam; this is freedom for both the artist and those who are party to the works that they produce.

It is not necessary to launch one's mind into the future to bring a liberating perspective to the space environment. Restrictions on freedom of movement, even with the best engineered spacesuits, will be a fact of life, and artists can take their viewers and listeners on vicarious journeys, perhaps inspiring them to overcome their lethargy, don a spacesuit, and head out on a walk, or perhaps a gruelling Trans-Mars Expedition, circumnavigating the Red Planet over both poles with an ascent of the summit of Olympus Mons to boot.[7] In this capacity, to render journeys across forbidding terrain, we see the artist as an explorer and traveller, a function to virtually release the population from its internment. Even if those who partake in this artistic exercise cannot follow those same exploits in the flesh, they can allow their minds to travel across these frontiers. These images may be similar to Earth-based paintings of ships on the high seas, plays depicting adventures to distant mountains, or even entirely fictional accounts of travels to lost valleys, the centre of the Earth, and other destinations far removed from reality.

Beyond imagining journeys, artists might join expeditions across the Solar System, painting as they go or returning to the safety of their habitats to render music or write theatrical works inspired by their expeditions. The writing, images, poetry, music, and plays of travel become forms of surrogate escape from one of the extraterrestrial environment's most tyranny-causing effects: the curtailment of the freedom to move and travel outside the confines of one's habitat.

In all of the foregoing functions, the style and excellence of the product will depend on the quality of the artist. In every age, we find those few painters, sculptors, writers, and actors who display what we generally refer to as genius. We cannot say what that quality is, and it cannot be easily distilled; if it could be defined exactly, maybe we could easily recreate it. But something inheres in their work that brings to mind brilliance. Perhaps it is the unexpected, the clever, or the nuanced that inspires in us a sense that they have seen deeper or gone further than anything before. Regardless of its source, these few individuals bring into our society the function of the artist as genius. More exactly, the purpose of genius is to elevate and give high expression to human activity, to awe us all with the capacities that our species can achieve. In the humdrum extremities and the monotone features of extraterrestrial environments, the rare genius will enliven the population to the beauty and exceptionality that the human mind can bring into life-destroying extremes. Because of its ineffable qualities, we cannot commission an artist to be a genius in the same way we can commission them to be a raiser of the spirit or to be an explorer and traveller, but where it does emerge, it performs one of the most exalted purposes of artistic expression.

The functions that I have described might be thought of as remote, detached contributions to society in that they are about raising the inspiration of the people, but they do not necessarily have, nor do they attempt to have, a direct impact on the institutions of the polity. As an extraterrestrial society emerges from the rocks or emptiness of space, so new organisations will form, and in this realm the artist has a role to play as the germination point for new institutions. This function is not unique to space; every art academy and institution that exists on Earth is an example of how artists have been nucleators of institutions, but in space, where humanity will start with nothing, a new opportunity and vista for this contribution opens up.

The idea that artists might act as a focal point for new types of collectives is not a purpose for the future. Already, such institutions exist. The Swiss-based OURS Foundation has flown art to space stations, and its explicit purpose and raison d'être is to use outer space, through art, to raise humanity's awareness and create new opportunities for enrichment and advancement on Earth and in space. On the supranational level, the International Association of Astronomical Artists is an organisation that has gelled around a group of people focused on depicting space, including its human exploration, in its many and variegated forms.

We can expect this associative activity to continue into the space frontier with lunar and Martian art academies and societies, even if they are comprised of no more than one or two people in a small lunar settlement who like to paint, play guitar, or pen plays; these individuals are still instituting new human arrangements that may later grow into full-scale organisations.

In this capacity, the artist is bringing into blossom new branches of the human political community beyond Earth, but more than this, the flowering of artistic institutions acts as a counterpoint and balance against the tendency for practicality on the space frontier. The cold technicality of the environment and institutions needed to live successfully beyond Earth should not be ignored, but their tendency towards drabness and hard calculation can be softened by the creation of institutions whose sole purpose is to encourage and support the production of art. The notion of a parallel activity should not be misinterpreted as meaning a separate stream of human activity, since as I have mentioned before, technical artifice, particularly in the matter of architecture, can be infused with artistic design. Artists not only balance functionality, but they can also actively work with those people charged with ensuring the technical competence of an extraterrestrial settlement to bring it to life and round off the hard edges of mechanisation.

In creating new institutions and, through those institutions, having a place to associate and share ideas, we have at once the emergence of another one of art's most useful societal functions: the artist as a norm-breaker. Artists have traditionally been norm-breakers in a multiplicity of ways. They might make objects that cause a frisson of controversy in society as to the value or artistic merit of the exhibit, thus challenging society's mores and conceptions of the creative. The persons themselves can be challenging, in the way they dress, the opinions they express, or their view of the institutions that surround them. It is not the case that this role is unique to artists; scientists can be eccentric norm-breakers as well, as indeed can anyone, for that matter. However, because the artist is expected to be 'creative', society has always given latitude and perhaps more free rein to the oddities and deviations of artists.[8]

In the extreme conformity-encouraging conditions of space, the artist as a norm-breaker can help intensify and bring to attention the spirit of individuality to an extent that is not so essential on Earth. That is not to say that artists should in a shallow, contrived way make efforts to be different for the sake of it, but where they genuinely wish to express views and styles that are in conflict with the accepted norms, they should pursue their role as norm-breaker with alacrity. The efforts needed to survive in potentially lethal conditions in space will give encouragement to authorities and the opinion expressed by the majority to suppress extreme displays of norm-breaking, but this very tendency is the reason why challenging convention takes on a special meaning in space and it behoves an extraterrestrial society to find ways that allow latitude for this role of the artist, and to encourage a culture that supports this.

Moving beyond breaking norms and the focus on the creative power of art itself, art is a commodity. Throughout history, artists have sold their works to make money, some to selected individuals, such as wealthy families or the church, some to the mass market. Art finds its way into homes, hotels, public spaces inside and outside, and in every place where humans collect; this generates an enormous market and insatiable demand for all forms of art.

In the earliest settlements in space, art is another form of industriousness to add to the diversity of ways in which the society might generate economic activity, both to bring into being and energise its domestic market and to provide material that might find its way beyond the settlement, perhaps back to Earth. I have had occasion to mention Alan Bean and his Moon art. Not only does it act as an enhancer of the spiritually uplifting influence of art, but it makes money. Paintings that have lunar dust sprinkled on them or a stamp of one of his Moon boots, or sculptures he made with similar tools used to do geology on the Moon, command a good price. In this function, we see the artist as entrepreneur, producing new products for society but in the process generating new capital markets of benefit to the movement of money, the health of economic activity and, with it, the breadth of economic liberty in the settlement. In space, artists should be encouraged by seed-corn economic or material support to engage in the entrepreneurial dimension of art; perhaps new organisations might be established to sell artwork on Earth. The achievement of this activity will itself assist in inculcating in the authorities that run the settlement the desire for economic and artistic freedom, lest their settlement take on the hue of an artistic backwater.

These functions of which I have spoken were those that the exhibition The Artist brought to my attention, and I have expressed these purposes in a way that emphasises their value in space. In almost all of them, we find that the space environment intensifies the value they might have in enhancing the conditions for liberty or subduing the detrimental societal effects of the deadly extraterrestrial environment. However, there are four other functions that I wish to elaborate on that are not necessarily unique to space, but that were not part of the exhibition. Upon reflecting on the role of the artist in society and melding it with an interest in space settlement, I think that these may be meaningful additional roles to which the artist might give their energy.

We have already seen how the artist can, in a variety of ways, act to liberate the minds of others, whether releasing spiritual urges or the desire to travel, or just breaking some norms. However, given the conditions in space, we might bundle many of these facets together to emphasise more generally the role of the artist as promoter of liberty. I mean to suggest that artists should go beyond using this function to encapsulate some of the purposes that we have spoken of already; they might consciously understand the obstacles to creating a free society in space and use their media to advance the spirit of liberty.[9] They might write satire, draw paintings depicting uplifting scenes of voluntary association of people, and so forth. As with raising spirits more generally, this should not be sickly, over-egged, and contrived, but the artist as promoter of liberty is a person who takes any opportunity to bring forward the values and spirit of freedom in a valid and artistically meaningful way.[10] They see themselves as a person with the power and authority, just like engineers and scientists, to contribute directly to the construction of the freest society possible.[11]

The practical insipidity of the technology used in space is evident in almost every image we see of humans in space. The International Space Station is surely a home to the astronauts that live in it, but I would hazard a guess that if you asked any astronaut whether they would like to live in a home that was laid out in a similar way for the rest of their lives, they would answer in the negative. This brings me to the next function: the artist as engineer and product designer, working alongside engineers to subdue technological blandness.

Artists have already played the role of product designer. The calibration target for the Beagle 2 lander, a European Space Agency lander that put down on Mars in 2004, was designed by artist Damien Hirst. Since he was famous for his 'spot paintings', the small circular colours needed to calibrate the cameras on Beagle, to make sure the images matched the real colours on Mars, were the perfect medium for expressing his skills. The calibration target, 16 small circles of different colours arranged in concentric circles, was both a functional tool of space exploration and a work of art. Beagle 2 landed on Mars but did not unfurl correctly, ending the mission. Nevertheless, within its structure is the function of the artist as product designer and engineer. As with all functions of the artist in space, this too contributes to the milieu of liberty by generating variety within engineering structures whose practical uses might well cause them to converge, like convergent biological evolution itself, into similar forms, thus contributing to an environment of conformity. Here we see how, by contributing designs and artistic flair to items of engineering, and by actively taking part in product design, the artist is addressing an issue of social magnitude in the extraterrestrial environment.

Throughout history, the artist has been historian, capturing moments which, despite being 'only' art at the time of their conception, later become historical records. There are innumerable examples of works of art that are deliberately historical, and we can count within this pantheon of work portraits and literary historical accounts. Yet in space, the artist might not only act as a historian for that particular location, rendering in images, written work, and music, depictions of a settlement at a particular point in time, but they might take on the role of terrestrial historian, bringing to the settlement reminders of the planet from which they originated. Plays and pictures and music and mayhem on stage may be the funnel to bring the history of Earth to the settlers, perhaps to remind them from where they came, perhaps to inform young minds unacquainted with Earth. If a settlement was to achieve independence from Earth, becoming completely materially self-sustaining to the point where it felt an urge to define its political and economic independence, then it may still wish to find a way to keep Earth, where most humans will live for a very considerable time to come, in the forefront of its cultural and historical awareness.

Located in the extremities of distant places, isolated from other nearby settlements, it may be easy for people to become jaded, morose, depressed even, about their surroundings. Not only is this state of affairs deleterious to the human mind and mental health, but it may play into the hands of those who seek dominance in settlements and who go about achieving this very situation by 'rescuing' those who have fallen victim to this ennui. Here the artist may find a function as a reminder of purpose. Through works of art, the population, even individuals, may be reminded of what they came here for: as a miner, as a scientist, as an engineer, as a tourist, or as an artist.

There are a veritable cornucopia of reasons for people to venture into space, and through music, writing, and images, artists can remind the population about the high-minded ideals, even in the mundanity of a tourist adventure, to lift their minds. In some sense, we find similarity with the purpose of the artist as enhancer of the spiritual environment. However, this function is not about a generic interest in making the environment more cultural and aesthetically rich, but specifically about reminding those far from home what their aspiring purposes were in taking that trip. It is a function that is not without use on Earth, where the artist may be a community member in an isolated population bringing life to their purposes, perhaps an Inuit hunting image; but in space, the sheer vastness of the spatial scales brings to the fore this function and gives it an energy, vitality, and essentialness that has no parallel on Earth.

It is no coincidence that dictatorial regimes have often suppressed art and artists as soon as they gain control over a national state on Earth. All of the functions I have explored here involve expanding the human creativity and free capacities for thought. They inherently challenge what we assume, alert us to the environment around us, and often invite us to envisage new pasts, presents, and futures. This essential effect of art, more than merely its existence as a thing of beauty or sublimity, is its contribution and legacy to liberty. In space, as on Earth, these characteristics will play out, but as I have sought to show, in almost every function that one can imagine for an artist, the effect is magnified or given greater urgency by the conformity encouraged by any extraterrestrial environment.

Artistic flair and expression, as on Earth, may well seem a threat to those who seek to impose regularity on a society faced with instantaneous death. But the artist is a most vital tool against this very tendency. With engineering, science, and education, it forms one of the four pillars of the structure of liberty, or another buttress against the encroachment of the tyranny-prone extraterrestrial environment into the human mind. We could just let artists run free and do their thing and hope that their creative juices will lubricate the free machinery of society; I think that is a very good idea. However, we could do what I have attempted to do here, which is to understand more formally the functions of art and artists in society and how those functions, and new ones besides, might actively improve the conditions for freedom. Through such categorisation, we might be better placed to direct support, whether moral or even financial and material, towards those ends and thus advance, in a most purposeful way, the functions of artists in space. Through such activities, we may not only improve the diversity, complexity, and richness of artforms beyond Earth, but in the process we have a chance to allow art its fullest expression in defeating the forces that pull society towards tyranny.

8

EDUCATING THE FREE CITIZEN

Education is a vital antidote to extraterrestrial tyranny

Francis Bacon famously observed that 'knowledge itself is a power'.[1] His terrestrial invocation is actually a rather elementary elaboration of the truth that if you have knowledge, you can better understand what is happening around you in the universe, perhaps comprehend it, and, most importantly, engage in discourse, disagreement even, with what you observe. His aphorism would be given particularly powerful impetus in a place where people could control the technology that allowed you to exist from moment to moment. The scientific knowledge behind that technology, the principles and laws that make it work, is wisdom essential to be able to apprehend and explain the functioning of the mechanism. On the frontiers of liberty, where freedom's range of operation is more easily delimited, education is an inestimable tool that allows the individual to judge decisions on the exertions of power.

Consider the act of an authority moving an individual from one habitat to another, the new habitat perhaps being substantially less salubrious than the person's existing home. The reason for the move is based on the claim that the existing habitat has a dangerous fault and that the person should be glad the governing power has stepped in to safeguard them. It so happens that this individual has recently criticised the settlement authority in a very public way. Are they being moved as a demonstration of power over them and to downgrade their living standards to boot, or is this coincidence? Perhaps there really is a technical glitch with the habitat and the individual should be grateful that their recent vocal calls for political overhaul have not lessened the authority's willingness to protect them. There is no grey area here: either the authority is being venal or it is showing a high level of impartial political integrity. To assume one, when it is the other, is either dishonourably paranoid or dangerous. Now, to challenge this decision, the individual must have a channel through which they can do this: for example, a court or a mechanism to demand a hearing. I would contend that such mechanisms should exist in a settlement. But even given these channels, it is necessary that the complainant and others around them have considerable knowledge of the technicalities of the settlement to be able to make that challenge with confidence and correctly in the first place.[2] Even though the individual might call upon engineers to help them make a judgement, it requires some basic knowledge to be in a position to know whether calling in assistance to challenge the decision is even reasonable.

We might also note that if an individual has the requisite knowledge, the authorities may be less willing to use their power coercively, being cognisant of the fact that it may be transparently seen to be unscrupulous. Thus, quite apart from technical information, we could underline the point that education contributes more generally to the capacity of individuals to engage in the polity. That need not be science education as such; it can be a general level of literacy. The technological edifice aside, if we want to secure democratic systems of deliberation in the extraterrestrial environment, we want to create an education system that can work at any age to offer sources of knowledge, critical thinking, and an understanding of governance.[3]

Earlier, I drew attention to the fact that no systems of checks, no perfectly designed constitution, no rationally designed order in a free society can secure liberty if the people are not receptive to freedoms and willing to defend them in their everyday interactions and behaviours with one another; the desire for liberty must reside in the minds of people. Education is vital to achieving this virtue of purpose and manner. It is essential to emphasise that I do not view education as a means to attempt to perfect, or achieve perfection, of the human character. Rather, education is a means to channel the pre-existing raw material of the human disposition in the most positive ways possible and to afford an individual with the knowledge to help achieve that end.

We might therefore recognise that education in a space settlement, whether that involves children or adults, is another crucial part of the edifice of liberty, and we might try to elaborate on what that education could be. In a speculative flurry of extraterrestrial pedagogy, this is a task I will touch upon in this chapter.

Education will be undertaken via a number of diverse methods

Much is made of the *means* by which we should educate people. This is not a book especially focused on pedagogy or predicting the scientific and technological capabilities of future schools, but, nevertheless, we might say something about possible styles of learning. In whatever way that we might imagine the human brain can potentially gather information in the far future, I suspect that much knowledge will be gathered by people through verbal and written communication, as it is today. Conservative though such a prediction might be, I see no reason why the time-honoured method of congregating in classrooms to learn in groups would not be a prominent method both for adults in continuing education and for youths.

Even if we can transmit information directly into the brain via a neurological link,[4] the social attractions of learning with other humans, and meeting new people in the process, have every likelihood of encouraging group learning activities. For youths, the chance to acquire social skills, even if the technological means to establish knowledge in the human brain have dramatically advanced, will surely mean that collective learning methods, even old-fashioned schools, will be demanded and created.

Furthermore, I am not a neurobiologist, but I am sure I know, in common with the reader, the importance of information 'brewing' in the human mind; this book is like a pot of tea that has been brewing for a few years. Even if we can neurologically 'inject' information into the brain, the process of discussing, interacting, testing and kneading ideas with other human beings is an essential part of the maturation and synthesis of knowledge. So I am not convinced that the proverbial ancient gathering of hominids around the camp fire will ever be extinguished as a means to learn.

I also see no reason for dogmatic approaches to designing how this education is to be delivered. Given the need for basic knowledge for a person to operate successfully within a settlement and contribute to survival without posing a threat to fellow citizens, it would seem sensible that the settlement construct some organised educational activities and curriculum—what we would recognise on Earth as a state-planned educational system. I will have more to say shortly about what such a curriculum might contain. Beyond that central provision, where individuals or groups can offer further education that does not represent a critical 'make or break' set of knowledge for the very survival of a settlement, I see no reason why private educational structures should not exist to expand on the diversity and type of educational opportunities on offer. These initiatives could range from courses in particular skills to something as simple as private tutorials from an individual who has some knowledge or proficiency to offer.

I have established a wide range of challenges to human freedom in space. We might, then, turn to the question: what should people learn to equip themselves with knowledge of how to successfully live in the extraterrestrial environment? Beyond that, what knowledge would be useful for them to develop their capacity for analytical thinking, the capability to confidently challenge the power structures that surround them in order to advance the state of their society and be aware of, and alert to, the generally tyranny-prone nature of extreme extraterrestrial settings? My intention here is not to devise a detailed curriculum for an extraterrestrial school or institute, but to focus on some general possibilities to embolden the conditions for liberty. In the following sections, I want to suggest some non-exhaustive ideas about the sorts of education that could be useful for the construction of a maximally free society.

Science and art, and particularly engineering, should be core education requirements in space

We could begin by observing, based on the comments that I have made, that an education in science and technology is essential, and that by maximising this knowledge, the degree to which the individual feels surrounded by mysterious technical wizardry, and the scope for the abuse of this lack of expertise by those in power, can be minimised.

In practical terms, this means that individuals within a settlement should have available to them educational opportunities to learn the basic science and the technical details of the machinery that surrounds them.[5] Those with formal scientific and engineering training within a settlement might be given the task of providing lessons and hands-on classes to the whole settlement on these matters. This education need not be compulsory, but the benefits of more fully comprehending one's environment would hopefully be self-evidently desirable to those who can potentially gain from these opportunities. This educational structure could be created from the inception of any settlement for an adult population.

If we imagine the presence of children in settlements, then this same education may be introduced in conventional, but compulsory, science and engineering classes. In extraterrestrial environments, this tuition must be more forcefully aligned with the knowledge required to apprehend the conditions to be found there. Thus, although we can imagine basic science classes as experienced in typical schools on Earth—for example mathematics, chemistry, biology, and physics—we can also envisage engineering classes with a focus on oxygen, water, food, and power production to establish in individuals a keen and early grasp of the mechanisms that keep them alive. Some of these areas of knowledge, in particular in the matter of oxygen production, have no corollary on Earth and would be distinctly alien. In essence, I am suggesting that space engineering would be a core curriculum requirement for any person in a space settlement, not merely to infuse them with good scientific and engineering prowess but, for the reasons I have identified, to give them the tools of liberty.

Recognising the sapping effects of the space environment on the expression of creative capacities that I discussed in detail in Chapter 7, these observations spill over into other areas of education, and they encourage a view that art, literature, and activities that encourage creativity will be of great use. For adults, continuing education in a variety of art practices will discourage the mind from being dulled into a petrified conformity by the extraterrestrial environment. This is no less true for young people, for whom art and its creative forms has an essentialness for preventing the emerging mind from being similarly stymied; by offering their youthful creative capacities free rein, we can encourage them to learn a social culture that finds artistic expression and the projection of one's point of view as acceptable, both in themselves and those around them. Put simply, they can learn to value the capacity for freedom of expression. Equally saliently, sessions in art will interpose themselves into a young mind's everyday experiences of the monotone colours, smells, and sounds of the extraterrestrial environment.

Brief periods of artistic flourish train the mind to revel in this opportunity for railing against conformity, and the intellectual effusiveness of the settlement might be energised through this training. To formalise some understanding of these tempers and influences, we could adopt the suggestion by de Vigne[6] that there would be education in Theory of Mind, which would introduce individuals to some philosophy and neurobiology concerning the functioning of the human brain. This erudition would enable them to grasp some of the precepts and theories that lie behind how we think the mind works and adapts to its surrounding conditions, inspiring in learners some self-awareness of the way in which the extraterrestrial environment shapes their perceptions, thus linking artistic and scientific insights through the interface of neurobiology.

The extraterrestrial situation encourages an educational 'trinity'

I want to go one step further into the more speculative realms of suggesting three areas of education that generally, at least as far as I know, are not systematically taught together in any schools on Earth, but I think recommend themselves in space. I will go as far as to say that I see no reason why they should not be very useful on Earth. However, my objective in this work is not to propose an overhaul of education systems on Earth. Whatever the generic usefulness, I suggest that these ideas have specific bearing on the conditions for liberty in space.

There are three areas of education which I collect under the generic title of a 'trinity' of education subjects (Figure 8.1). These three areas I call Methods of Life, Liberty, and Planetary Knowledge. I will explain the rationale and basis behind each of these

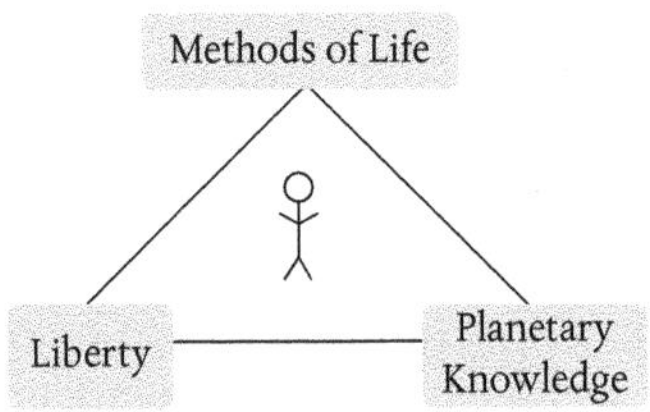

Figure 8.1. The education trinity. The education of a free and virtuous extraterrestrial citizen can be partially accomplished by a trinity of subjects that cover the essential capacity to objectively critique and analyse information about the physical and human world (Methods of Life); the skills to understand and implement the arguments and methods that secure freedoms (Liberty); and a comprehensive scientific understanding of the planetary environments that define the types of social conditions in which one lives (Planetary Knowledge). This trinity is embedded within other educational strands including science, history, and basic engineering.

parts of the trinity in turn. Each augments the traditional academic subjects by not only imparting knowledge and facts to the individual, but giving them an improved capacity to use that scholarship with more effect during their lives, in particular for the building of successful free societies.

The trinity, part 1: Methods of Life

If the inhabitants of an extraterrestrial settlement are to successfully understand the scientific and technical basis of their existence at the edges of physical extremes, they must have more than a grasp of the essential facts and knowledge of the matter: they must have the capacity to understand the methods by which we assemble evidence into viewpoints and argument—in short, the scientific method. I hesitate to call it the 'scientific' method, since I think this way of understanding the universe—by assembling facts, testing hypotheses or ideas about how those facts hang together, and finding evidence that falsifies those ideas—is a very general method that recommends itself to all walks of life. It just so happens that it was scientists (or natural philosophers, as they were called before the nineteenth century) who developed the methods most forcefully, so we tend to call these approaches the scientific method.

To challenge an extraterrestrial authority and to be able to make judgements about the running of oxygen, water, food, and power production facilities and all of the health and safety demands that emanate from these processes, it is essential for a person to be able to assemble, analyse, and derive conclusions from facts and to be able to know how to drop ideas when good evidence falsifies them. Individuals must be able to debate the quality of so-called 'good evidence', and to refute conspiracies and hysteria. All of this falls under the purview of the history and ways of testing hypotheses described as the 'scientific method'.

Thus, I refer to the first part of this trinity as the Methods of Life. I deliberately leave the word 'scientific' out of it, to emphasise its broad use in all areas concerning the analysis and discussion of evidence to do with one's life.

For adults, this education could be delivered as continuing education. It would begin with an examination of ancient philosophies and the methods of those traditions, leading into the experimental histories and methods of Francis Bacon and then following on to Newton, Whewell, Jevons, Popper, and many others who have formalised the logic and methods of what we call the scientific method. With respect to the extraterrestrial case, I think the most important aspects of this education are: (1) to gain some knowledge of the essential logic behind developing hypotheses, testing them by falsification, and to understand the pros and cons of the various approaches proposed by those who have developed these methods; (2) to understand the role of probability and the lack of dogmatic certainty in any proposition about the universe around us; and (3) to be provided with examples from the real world with which to practise the use of these methods. This might include hypothetical decisions about oxygen production and the data and facts that must be adduced to conclude something about the usefulness of decisions.

The endpoint of this type of education is to develop rational critical-thinking skills and to understand the power, but also the limits, of enquiry and how this can be applied to a society in which science and technology play such a pivotal role in the everyday existence of settlers.

Methods of Life lends itself to varying degrees of complexity depending on the audience. Adults with some prior scientific knowledge might grapple with some of the formal logic of methods of analysis, such as those developed by Popper, or undertake ancillary courses in Bayesian statistics to understand the role of probabilities. A general audience might learn about hypothesis testing, falsification, reproducibility, and probabilities of truth embellished with some history of Bacon, Newton, etc. A young audience might be introduced to some history and some anecdotes about early science.

No less than for adults, crucial for youth is that instruction is given on the application of these methods using real-world examples, to hone the capacity for critical thinking. As an example, imagine a group of schoolchildren who are asked to derive an outlandish hypothesis of any kind—maybe that their settlement has an enormous deposit of gold underneath it. They must then develop the means to test that hypothesis, for example by drilling, seismic studies, organising an enthusiastic group of school children to dig, or other approaches. They must then consider how their hypothesis could be falsified. Would a single drill core from one location that failed to find gold be enough to falsify the idea? Surely this would not be sufficient; they must have many drill cores, and thus a discussion on replication, and the number of drill cores that must be obtained to achieve some desired probability of certainty in the result ensues.

The Methods of Life is not just a way to tighten critical-thinking skills for these sorts of engineering applications. In developing them, people learn how to better critique political and economic data, and indeed all data presented to them in its cornucopia of forms. Learning about the scientific method, its history, and how to use it is a liberating education, since it equips the individual mind with the power to question authority and the opinions of one's peers, and it enables one to fully engage with the debates and decisions of society. In essence, a knowledge of the scientific method is a gateway to individual liberty.

The trinity, part 2: Liberty

Another area of enquiry which I think has much to recommend it, and again I would point out would be equally useful on Earth, is the various concepts of human freedom. As we have seen, liberty has no easy definition. We might speak of political, economic, or various kinds of social freedom. Within the word is a maelstrom of discussion. There is the metaphysical question of what freedom is and whether the human mind is ever 'free' in any real sense—in its state as a concatenation of biochemical processes. Beyond that more fundamental set of ideas are the ideas that shape our society. What do we mean by liberty? How much individual freedom should people have, and when is it justified for people to be coerced into carrying out orders? To what extent is liberty the state of being free of domination or interference or the capacity to realise a set of rights? As the reader is aware, there are millennia of human experience addressing these questions, and centuries of published work.

Beyond Earth, surrounded by the paraphernalia of survival, a deep understanding of the arguments about liberty is essential if those individuals are to be able to think and argue with cogency about the decisions taken with respect to this pervasive technical infrastructure and the political and economic culture that emerges around it. If they are to cherish the institutions of liberty, uphold the delicate structures of democracy, and be vigilant to its threats, then everyone must be trained with a philosophical turn of mind, an interest in political philosophy, and some grasp of the literature that provides them with the lessons of the past, even if those lessons are primarily from terrestrial societies.

An education in liberty need not be politically skewed: one can read and discuss Nozick's *Anarchy, State and Utopia* and Marx's *Das Kapital* without having to take sides. There are a profusion of questions about freedom, including the obvious matter of what the word itself means, that admit of no required political view. What I perceive to be educationally crucial, however, is an understanding of these historical arguments and their origins. Many of the freedoms that exist in various nation states on Earth today, such as equality before the law and the rule of law,[7] are concepts and insights that have taken many centuries to develop to their full potential and application, yet a large number of people remain ignorant to the efforts, and in some cases sacrifices, that were made to open the human mind to the value of these ideas and their application. Freedom of speech, for instance, an apparently elementary idea, allows for complexity in its interpretation and history. There is no definitive statement of what it entails, and although there is some principle being elaborated when a person or group of people states a commitment to upholding freedom of speech, it is, rightly, a concept that is worthy of constant investigation and the true comprehension of which requires some work.[8] The free expression of ideas can, of course, harm some people, which explains the long history of debate about exactly what we circumscribe within the acceptable limits of freedom of speech.[9]

An education in the history and ideas of liberty is not only a means to gain an appreciation of how one's own society developed, but this historical and intellectual foundation is necessary to challenge and offer new insights into the never-ending process of debating liberty. The arguments that lie behind many of the political and economic ideas of freedom are intertwined with history, yet at the same time can, from some standpoints, be seen to be rational and independent of historical vicissitudes.

The concept of being free from the fear of arbitrary arrest, or execution, is a freedom that I think most rational human beings would regard as a social advance from the long stretches of history in which this condition did not exist, although sadly some nation states on Earth today have not reached this point. What forms of freedom might we regard as universal, objective even, and which mere reflections of specific cultures? To answer these sorts of questions requires grappling with both the history and the theory behind our ideas of liberty.

As with the Methods of Life, we might imagine that gradations of complexity could be created for different audiences. Young children might learn something of ancient Athens and Sparta and discuss and consider the different social structures of those societies, learning more anecdotal information that is appropriate to minds in the early stages of their analytical capabilities. Older minds would be opened to the full canon of work on freedom—Confucius, Kautilya, Grotius, Montesquieu, Locke, Hobbes, Kant, Mill, Gandhi, Marx, Rawls, Berlin, Hayek, ad infinitum—addressing the arguments in those works and debating and discussing their applications. They would use this canon of literature to consider the different forms of liberty—metaphysical, political, economic, civil, and so on—and to debate the very meaning of freedom in their own environment and others. They would touch on the systems of government in ancient Greece and Rome, reading parts of Aristotle, Polybius, Cicero, Sallust, and others, gaining classical insight into the early forms of monarchies, aristocracies, democracies, and republics and their successes and failures.

Essential to the assimilation of these ideas is their real-world application and practice, and here we might imagine animated discussions, particularly among youths. We could fabricate an imaginary decision to centralise oxygen production or make it mandatory for every citizen to spend two weeks a year tending to oxygen machines. Any number of hypothetical situations might be constructed to encourage discussion on the merits and demerits of such decisions with respect to the conditions for liberty. Is it right to extract enforced labour from a person? What about if it pertains to the production of oxygen on which the whole settlement depends, including the individual carrying out the task? What are the effects of such compulsory demands on the wider social conditions? How would one decide which was more consequential: to carry out emergency repairs on a failed oxygen machine that threatened the lives of ten people or the life of one person who is crucial to maintaining oxygen machines? From philosophical extraterrestrial trolley problems, with their implications for the liberty of people, to labour relations, endless arguments that are rooted in the history and ideas of freedom emerge from a single situation.

It is relatively easy to educate a person to express their freedoms. In some sense, this objective is a crude and blunt-edged invocation to self-realisation. However, as I have addressed, a much greater challenge is allowing healthy dissent in confined environments where people are so dependent on a number of vital supplies. Expressing dissent in a forceful, provocative, but non-destructive way is a skill. Civil disobedience and the culture of this form of dissent must be inculcated into a free society, and its citizens must be able to wield this weapon with discretion and respect; it requires education.

Surely, one purpose of an education in Liberty is to explore the history of dissent and its forms. Students might learn about blunt forms of dissent that led to destruction, including terrorist actions, and the forms of civil disobedience given expression by individuals such as Gandhi and Thoreau. What makes dissent acceptable or not? There is no clear-cut answer or set of criteria;

the point is rather to gain a sense, a judgement, of where that line might be drawn, and that division will be modified as society changes. A physically small settlement with one oxygen machine is likely to be less tolerant of acts of civil disobedience from the engineers who run that machine than a settlement where there is redundancy in the mechanisms of oxygen production. Surrounding that observation are the other multifaceted characteristics of a society, such as its population size, the extremity of its environment, and its ineffable culture and history. Yet despite the inexact nature of dissent, we can equip individuals with a historical knowledge of how humans have traditionally expressed dissent, the advantages and dangers of different means, and how those means might apply in the extraterrestrial environments. Through such efforts, we might educate a population to be able to better maintain a balance between the observance of custom and laws and the expression of healthy disagreement. Given this fine balancing act, we are seeking, dare I say, to educate a population that can dissent about dissent.

From such an education in the broad ideas and history of liberty, a sharpened appreciation of the conditions for freedom emerge, and concomitantly, a capacity to shape and develop arguments of benefit to one's active involvement in the polity arises. Liberty is a social art, not a crude demand for individualism and self-expression. Despite the caricature of classical liberalism (and the common foil that liberalism is a selfish creed), constructing a sphere of self-realisation in society while respecting others and contributing actively to the collective whole that sustains one's power to be an individual is a deeply complex, learned set of behaviours that few societies in the history of our civilisation have developed to a high level. Developing the subtle blend of possessing an assured self-determination and self-respect in the face of authority yet harbouring a deep sense of duty to contribute to the society that provides the space and respect for such attitudes is the sweetest expression of the word 'liberty'. The aspiration of a society to raise its population in these capabilities is a measure of its commitment to the excellence of human thought and social organisation.

Quite apart from deepening an appreciation of liberty, its history, and its application, we could also wonder whether formal training in Liberty might make people less liable, on the whole, to seek the power of arbitrary authority themselves. This is a matter for psychologists to address, and I do not know if learning about the long struggles for the most basic forms of freedom would make a human more alert to what is to be lost when they find themselves with the opportunity for power. Christian Bay states with force:[10] 'The better the child has been enabled to preserve his psychological freedom, the less likely is the adult to develop a taste for exerting power over other people.' This is more a statement about the treatment of a child than their education. However, I think one might make a claim that a society in which people have an extensive and comprehensive understanding of the literature and thoughts of liberty is at the very least one where a larger number of people are intellectually equipped to dispute claims for power, and that must surely be a good thing, as it probably is on Earth.

It will be a challenge of the highest degree to human integrity and education to build a society where the febrile structures of democracy, voluntary association, dissent, and human self-expression can be made real against the incessant clarion calls for control and predictability. Education could surely be a centrepiece of achieving such an order, and the design of a curriculum in liberty would surely recommend itself.

The trinity, part 3: Planetary Knowledge

I now turn to the final segment in the trinity. When one lives on a world such as the Moon or Mars, where the space around is an all-embracing deadly carapace and the environment is strewn with rocks more than a billion years old, one cannot escape a need to understand the history and function of that world, to know one's place in it and the reasons for the type of society that has emerged. Even in the free plateau of space, far away from life on the surface of a planetary body, an understanding of the worlds that occupy that space is essential for appreciating the distribution of resources and the formation of solar systems in which civilisation develops.

Before the reader rebukes me for suggesting that these statements are particular to locations beyond Earth, we can see easily enough that this discussion is admirably relevant to Earth. On a planet that has a highly productive biosphere incorporating vast rates of photosynthesis, a natural life support and oxygen production system, a critical understanding of how it works is essential to comprehending the nature and future of our civilisation. When Earth's biological diversity and atmospheric composition is subject to serious change caused by the industrial activities of a single species, then there can be little hesitation in believing that an education in Planetary Knowledge is of the utmost urgency.

I draw attention to an interest in Planetary Knowledge beyond Earth because in some sense it is the lack of planetary functions, such as photosynthesis, plate tectonics, magnetic fields, and so forth, that renders all planetary bodies other than Earth mortal to the human organism. To understand why one is immersed in a complex artificial life support system and the social systems that emerge from it, one has to understand the formation and the geological, atmospheric, and maybe even biological, evolution of worlds.

The study of how planets are formed and what factors determine their long-term historical trajectory leads to a comparative understanding of how planets do, and do not, host life and the multiplicity of ways in which those worlds may evolve over time. In the case of Mars, to understand the small size of the planet is to understand why its magnetic field shut down and why plate tectonics ceased early on, if it really ever got going at all; it is to understand why the atmosphere is thin, why there is no liquid water on the surface, and why the radiation doses are high; it is to understand why life there could never have emerged into a productive biosphere like Earth today, capable of producing oxygen, and why the landscape is a tapestry of reds, oranges, browns, and greys. Most importantly, it is to understand why Mars is different to Earth and how those comparatively different trajectories shaped those two orbs. We could make similar observations about the Moon and many other planetary bodies.

Planetary Knowledge is a scientific education, an interdisciplinary journey into geology, biology, geomorphology, hydrology, physics, chemistry, and more besides; but it is also to grasp why the social and political world is like it is. The lack of oxygen, water, and food and the consequences of that reality on the society that emerges is a function of environments in space, with the set of conditions in any different location formed and shaped by the history of that world or empty void. When we gain this insight, we can better understand why a lack of naturally produced oxygen is not a denial of liberty but an inevitable feature of existence beyond Earth, and that sources of illiberality must be sought in the social consequences of that lack of oxygen, not in the fact of the paucity itself.

In investigating the features of planets, Planetary Knowledge would include many of the aspects of geography education that we associate with terrestrial education. On Mars, for instance, learning about ancient meandering rivers and volcano formation would have its parallels to Earth, but we would imagine that a focus on impact events and their consequences, as on the Moon, would play a more dominant role in classroom discussion because of the prevalence of these structures on those bodies. Here again, we observe that a curriculum will be substantially transformed and modified from examples on Earth, yet not unrecognisable, since its subject matter is anchored in the same essential scientific principles.

To summarise, I have referred to these three areas of education as a trinity because they are linked into a synthetic whole. A grasp of the methods of sifting, analysing, and critiquing data is an essential component of comprehending the practicalities of realising the liberty to express ideas and allowing others the same. A scientific understanding of the history of a planetary environment is essential for knowing which decisions are an ineluctable part of the environmental conditions in which one resides and which are politically motivated, and so forth. Although split for convenience into three segments, they seamlessly grade into one another as a human navigates the scientific, political, and economic environment in which they exist. A wide-ranging depth of knowledge about methods of thought, liberty, and one's environment are an intellectual triumvirate upon which much else can be built.

It is obviously the case that to deliver this curriculum, a settlement would require teachers who have the educational background and knowledge of these subjects to be able to promulgate them. In an early settlement, which will be numerically small, much is being asked of them to garner the full breadth of this proposed education, and some teachers may well possess other vocations. As a settlement expands, so specialisation delegated to different teachers will become more practicable.

The trinity I have proposed here is not a new-age dogmatic attempt to create a novel type of education. My purpose is to draw attention to the educational consequences of extraterrestrial environments. The extraordinary challenges they represent suggest that certain types of education could be of value to survive them, to understand their influences, and thence to build societies that are more effective within them. As my tract focuses on the conditions for liberty beyond Earth, I have made suggestions for certain types of educational focus that might enhance awareness of this matter and improve the ability of societies beyond Earth to dampen the chances of arbitrary and tyrannous influences, and relieve them where they do emerge. As always, the most effective way to create societies in which individuals are respected and the fine balance between collective survival and individual aspirations can be successfully maintained is to have an educated population. I have little reticence in pointing out that there is much more we could do to define educational possibilities, methods, and curricula that would enhance the success and survival of free societies in space.

On the matter of history education

Since there will be no long stretches of prior history, at least in the early stages of settlement, we might wonder whether the traditional subject of history education has any value to settlers. As I hope this work has made clear, terrestrial societies have much to teach us about living within the rest of the cosmos. History is that great unpredictable vortex that mixes cultures, ideas, and human action, yet exhibits patterns and principles, the eddies and swirls from which we can extract clues to improve our future.

Although a collection of information on terrestrial history and the various nations that have existed on Earth may be of some interest, it could be so expansive and disconnected that it would be difficult to make sense of it, or to make it cohere, unless it was moored to some common theme. The education in Liberty could be fruitfully used to provide such a foundation for lessons in history. I do not think that history lessons should be politically motivated; on the contrary, they should aim at a rigorous teaching of the facts. However, if we are in search of some thread with which to hold together a discussion of terrestrial history, the forms of government that have developed on Earth and the types of societies that they have presided over would seem to me to be an imminently sensible basis for ordering history education, and a theme that would emphasise an aspect of terrestrial history that has immediate relevance to extraterrestrial learners. Indeed, it is difficult to see how an individual can construct a credible intellectual foundation for liberty, even in the extraterrestrial environment, without a grasp of the history of societies on Earth.

Such a focus could take students from hunter-gatherers through to Neolithic civilisations of the Fertile Crescent and on to ancient India, China, Greece, Rome, the emergence of the nation state, and the structure of governments in more recent times. In studying the different societies, students might then take tangents to consider the lives of these people, the challenges they and their types of society faced, and all the other material that we associate with the teaching of history. In other words, the study of human organisation and its forms would be the guide rails of history education. This historical knowledge may then motivate the learning of certain terrestrial languages by some students. Of course, as time progresses, history may become increasingly focused on the developments that have occurred in the students' own location, but these more local developments could also be discussed and taught through the lens of social organisation and its history by making comparison to societies on Earth, ancient and modern, and deliberating the lessons that have been learned.

On the matter of religious education

The reader might have wondered where religious education sits within this outlook on extraterrestrial education. Religion, at least with respect to the existence of entities that created the universe, or continue to maintain it, is, in my view, a very straightforward binary matter. Such entities either exist or they do not. This is a thing to be discovered—or not, as the case may be—and the evidence is to be discussed in the same way that individuals might go about debating any phenomenon or conjecture of interest in the natural world. From this elementary view, the non-involvement of the governing authorities in religious affairs, and the prevention of those who believe in the existence of such entities from controlling schools or other social domains for the purposes of advancing their particular hypotheses, would follow. Just as it would seem odd for an enthusiast of Dark Energy to set up and run a school based on this one hypothesis alone and structure the day around this single subject, it would be strange that anyone would set up a school or an educational focus within an establishment around the existence of gods, not because of any perceived irrationality in the belief in the existence of gods as such, but because it is an unjustified and distorted focus on one specific matter concerning the nature of the universe. The approaches that we might use to test the hypothesis of such entities are circumscribed by the scientific method and indeed, among many other possible examples that could be used as educational material, a critical discussion of a God or gods could be satisfactorily incorporated within Methods of Life.

Although this may elicit horror in some people, my view is that religion should be covered in science classes: not because I wish to see the rigour of science distracted by apparently intractable problems, but because the existence of gods is a scientific question and the proposition that they do exist is a hypothesis that must admit of some form of testing. If the claim is that this question does not yield to such analysis, and gods can never be revealed by any methods available to humans, now or in the future, then it becomes an objective discussion on the limits of analytical methods and how such uncompromising boundaries, if we think that they truly exist, might shape our understanding of reality and the universe. Therefore, either way, matters of gods must always lie open to rigorous critical evaluation, and they lend themselves to the undogmatic discussion of evidence, its quality, the means by which we fathom the nature of the universe, and the limits to those methods. On the matter of understanding the history of religions and their various origins, manifestations, and activities, I think that these can be adequately covered within conventional history classes.

Likewise, morality or personal views that might emerge from the belief in the existence of gods or any other forms of 'higher' entities is not a matter for formal education. It would be as if I changed my behaviour towards my fellow humans, or took on customs and behaviours of certain forms because I believed in String Theory. No one would deny me the freedom to have those feelings and beliefs, but they would be my personal concern and not influences that should shape others, unless they felt so inclined by free choice to join me in my beliefs. Thus personal views, beliefs, customs, traditions, and behaviours that emanate from religious outlooks are things to be treated as personal choices, not the formal foundations of educational curricula, which should rather seek to equip the individual with the tools and capacity to intelligently adjudicate and critique these choices in the first place for themselves.

Thus, despite the supposed complexity and considerable amount of controversy that is sometimes attached to matters of religious thought and education, my personal view is that there is no reason why they cannot be dealt with on the basis of simple guiding principles that satisfactorily disentangle the objective matter of the existence of higher entities, which is a scientific question, from the personal matters that arise from these conjectures and speculations, which are indeed personal matters.

On the characteristics of a virtuous extraterrestrial citizen

At several points in this work I have touched upon the observation that liberty cannot be assured with constitutions and political structures alone; it lives in the minds of people, and they must be willing to nurture and defend it. We might wonder, therefore, about the characteristics of a virtuous extraterrestrial citizen in which these traits take form.

I do not think that an extraterrestrial authority should take on the responsibility of moulding citizens to certain specifications of personality, as this is to invite another veneer of conformity and the use of the production of 'virtue' as an excuse for wide social interference. Nor, as I have emphasised for the religious case, should it seek to enforce certain views about the way the universe works. Despite that, I do propose that certain character traits, values and outlooks can improve the chances of success for the individual and society in extreme settings and, importantly, nourish the conditions for liberty. The means of education and upbringing and the wider culture of society might encourage these characteristics and bring them to the attention of the citizens as traits worthy of their full and free consideration. Here, I wish to suggest what some of those attributes might be, since they link directly into the educational environment that we create.

A virtuous citizen in the extraterrestrial environment is one who understands the tendency towards the expansion of power into a greater domination and interference over the individual in an environment where conditions may threaten instant death, but who, at the same time, grasps that those same conditions mandate an effective social cohesion. Cognisance of this balance is at the heart of everything they do. This insight they can gain through experience, but it can be strengthened by a formal education in liberty.

Given the potential in these environments for mistakes that could have far-reaching consequences, such a person displays sympathy, forbearance, and forgiveness, particularly when they see error in others. They eschew belittlement of those who err, but rather seek to help and improve both themselves and others in these qualities. In severe emergencies and crises, one cannot find a person who will stand more solidly alongside even their worst enemy for the survival of the settlement, a characteristic rendered real by their magnanimity. Through these attributes, other individuals benefit and social safety is advanced, and through both of these, in turn, they themselves become a more selfless, responsible, effective, and meritorious citizen.

The virtuous extraterrestrial citizen is one who expresses their individualism with enthusiasm in such conformity-inducing extremes, realising the necessity of this for developing freedoms, novelty, and solutions to society's multifarious problems and challenges, but their personality is infused with a serious and cultured respect for the boundaries of safety, the security of others, and the need for that character to sit within the success of the collective that is confronted by precariousness. In other words, they exude a personality of individual self-expression surrounded by a bodyguard of social conscious and awareness. In appreciating the same aspect in others, they do not criticise, even with a sour stare, the personality and chosen expression of others. Such a citizen relishes, and takes great delight in, beginning new activities, and expressing ideas and views, that exhibit an unabated and unyielding individuality and self-determination, yet they prefer projects which, in parallel, produce products and outcomes that are seen to give direction and expression to the wider social project of securing society and the liberties it protects.

Extraterrestrial virtue is expressed in the highest degree of probity and honesty in conduct. Dishonesty, when embedded into a culture set in extremes, will metamorphose into distrust of all actions, which itself will mutate into paranoia and fear about actions that can easily threaten lives. The individual, then, has an important role to play in enriching this general culture of integrity, a rectitude bolstered by a controlled self-interest, a concern for the long-term commonweal, and by a personality not taken to flattery, sycophancy, or offers of power. Although they may crave power, for they are only human, they relentlessly keep these sentiments cowed and beholden to rational action in the cause of the democratic republic. This trustworthiness is to be found in their political, economic, and financial interactions; they never encourage the growth of shady deals, conspiracy, subterfuge, embezzlement, and other maladies that are so lethal to the body politic in extremity. In all matters, particularly those that deal with issues of considerable public safety, such as oxygen production, their uprightness is not only completely failsafe, but they see to it that those around them command similar public trust. In this same regard, just rules and laws established by democratic consent command their deepest admiration and esteem.

The virtuous display an egalitarian sense in their treatment of others. At the limits of human existence, the baker, the plumber, the oxygen manufacturer, the scientist, the artist, and every other worker has a role to play in taking society beyond mere survival into a state of social success. A virtuous person grasps this essential reality and this is reflected in their treatment of others. They do not hanker for a communist equality, recognising that inequalities emerge naturally through different dispositions, skills, intellects, physical capacities, and so on, but they treat all citizens as equal in the extent to which they merit basic human respect, whether they lie below or above another in any formal organisational hierarchy.

In extremes, where sometimes certain items may be in short supply, the virtuous citizen emits a generosity and humility towards others and an awareness of the needs of others, while demonstrating modesty in their own demands and moderation in their general conduct. Surrounded by instant lethality, the virtuous citizen is a serious personality who is hard-working but can face these realities with a sharp sense of humour and a sparkle of self-deprecation. Where they are wronged, and such situations are almost an inevitability in the course of a human life, they have a keen sense of forgiveness with an emotional control that keeps shackled any temptations to revenge or vengeance, emotions which, if encouraged across the social landscape, could be cataclysmic. Against the desolation of the universe, the smile and the look of kindly human eyes is the most valuable gift that one human being can bestow upon another, and this offering is without material or financial cost. The virtuous citizen is alert to these simple means of enhancing the social condition.

Given the vast technical edifice needed to thrive in these environments, the virtuous citizen is never one to shun learning, and anti-intellectualism is something unknown to them. They revel in having knowledge, particularly of science, but not in displaying that knowledge in a brash and superior way, even if they possess a better grasp than their neighbour; they generously sprinkle their know-how on all around them, while being the glad recipient of others' insights. This widespread embrace of proficiency not only protects the community but is the surest guarantor of the capacity of individual self-realisation.

Just as we have seen the importance of forms of art in countering the culturally sterilising effects of the extraterrestrial environment, we note that virtue also inheres in an enthusiastic support for these activities. The virtuous citizen does not cast dim eyes on works of art and expression, even when the most serious emergencies demand fealty to the immediate technical needs of their resolution. They never treat these pursuits as frivolous and trivial against the grander social horizons; rather, they take great interest in and express keen encouragement for art in all its forms. It is common for such a citizen to have at least one outlet for these activities themselves, whether it be playing a musical instrument, painting, writing, or some other activity.

Given the precariousness of liberties of various kinds when under the duress of the extraterrestrial environment, the virtuous take their education in liberty seriously, never taking for granted the effort, work, and sacrifices made by others to secure the sometimes meagre liberation possible. They read widely in political philosophy and enjoy wholesome argument and discussion about the means of achieving the betterment of society in this regard.

In places where the smallest error can result in injury or even death, virtue is reflected in the way in which the citizen places emphasis on the examination and discussion of evidence and the formulation of coherent hypotheses for the way ahead. In everyday matters, the citizen may well make small leaps of faith for the sake of expediency and efficiency, realising that a tedious over-analysis of every decision in life can make them not merely a bore, but a danger in the way in which they slow the capacity for action; but in the matters of consequence and moment in life, they place the utmost value on critical thinking, the acquisition of sound evidence, the demand for such evidence, and the ferment of discussion, argument, and hypotheses in coming to decisions and outlooks. They are immune to superstition and have no fear in discounting shibboleths in favour of carefully weighted evidence, no matter how much others may wish to cloak such rigour in the garb of sterility and overly scientific or cold thinking. They apply this attitude not only to the matters of everyday existence, but also to the loftiest thoughts about their ultimate purposes.

Through these character traits and others, citizens may create a way of living, a culture no less, that is conducive to freedoms and open deliberation among the citizen body. These virtues will make the extraterrestrial settler less prone to the vagaries of chance and fortune that can lead to disaster in extreme settings, and provide for them a greater possibility of the rational control of their lives and the outcome of their decisions in favourable directions. Education cannot secure these attributes; they must also come from training, upbringing, and other sources. However, education in matters of critical thinking, methods of scientific thought, and liberty can cultivate a philosophical, thoughtful, educated, and intelligent frame of mind in which these virtues have a higher chance of taking root.[11] Not least, in the literature I have mentioned, most particularly the classical literature—one might mention Cicero, Sallust, Livy, and Tacitus as four examples in the Roman pantheon that are notable in this regard—discussions on ancient republics were strongly woven into deliberations on what constitutes virtue.[12] Thus, in pursuing an education in liberty, it turns out that we gain, by default, a fortuitous means to deliberate how ideas on the characteristics of virtuous people have evolved over time and what assemblage of character traits constitute the virtuous extraterrestrial citizen.

It will be necessary to cope with the challenge of youthful exuberance

There is one specific challenge that is associated with young persons in space, and that is the expression of youthful exuberance, both physical and mental, in an essentially physically confined environment.[13] In essence, the problem amounts to the challenge of how to allow the freest expression of youths in an environment where people will tend to demand safety and predictability.

Youths in an environment where their superiors are alert and on edge in the face of unpredictability, and where parents do not wish to sire errant individuals, may be naturally subdued. We could expect that in the harmful conditions of space, young people may be more cautious and restricted in their behaviour; they might naturally possess a Stoic seriousness in their demeanour.

They will not have gambolled and rolled in the grass with their friends, climbed trees in limitless forests, and played hide and seek in quarries and caves; these activities will have been planned and contained under pressurised roofs if they have happened at all. Even if they wish to be highly expressive in a physical sense, the inability to easily project that in the outside environment may militate against it. As with my observations on art more generally, it might equally be the case that the confining peer pressures of extraterrestrial environments will be overcome by youthful expression in the mental sphere, in arts, and in written work.

Although a quite natural demureness may take root in the extraterrestrial child, we might wonder what to do about the potential for that unpredictable risk-taking, the prodding of physical limits and social authority that explores the territory of danger in confines. One controversial solution to the development of children in such environments during their effervescent and raffish younger years is the artificial control of hormones through some type of biochemical intervention. We cannot presume to know the ethical landscape of future space settlers. Certainly, in most places on Earth today, the suggestion that deliberate widespread hormone control in the population should be used to tame wild youths in places where damage to infrastructure is costly or potentially deadly would, I suspect, be resisted strongly. In space environments, a similar mindset might apply, yet perhaps for extreme cases, this may seem an ethical option for the control of miscreants in comparison to its potentially deadly consequences.

The question becomes whether education, parental and otherwise, can sufficiently control youth populations to prevent a requirement for some type of biological intervention in environments where deviance might kill many. We have not performed the experiment, but I suspect the answer is in the affirmative in the same way it is for populations of people who live in extreme Arctic conditions. Children are supple of mind and they quickly learn to intuitively grasp the culture around them. Instead of uncontrolled youths, it is perhaps more likely that children will rapidly master the realities of their confined condition and expecting nothing else, since that is the environment in which they were raised, a peer pressure for successful contribution to the body politic is as likely as errancy. Perhaps the confined conditions beyond Earth will breed, without purposeless effort, young people with a strong sense of their individual liberty expressed through their contribution to the polity.

Dealing with perceptions of Earth will be a challenge in extraterrestrial education and liberty

There are not many contexts in which Earth is described as a problem. However, it presents a challenge in the way in which we educate people living beyond it. Compare Mars, the most Earth-like planet in the Solar System, to Earth, and you can see immediately the source of this disquiet. Earth is a profusion of biota set in a landscape whose colours are limitless and which is infused with the sounds of wildlife, water, and wind. Magnificent oceans crash against sandy beaches; wind ripples through the leaves of autumn; the sun sets over a distant forest. One can walk freely and breathe the atmosphere, taking in the smells of flowers and settling down to watch the white water of a river or the glistening reflections of a lake. It is not difficult to see how Earth will seem like a paradise compared to any location in space to those who have moved away from the planet, even for a short while; this will especially be the case for those who are born beyond Earth and for whom this world is a set of diffuse images of distant verdant lands and azure oceans. Children in spaces beyond Earth will have been born in conditions that amount to cages, which is likely to give them a heightened awareness of any planetary environment which appears to offer seemingly boundaryless movement.[14]

Personally, I have experienced this set of emotions after spending relatively short (by polar science and exploration standards) three-month stints in Antarctica and stepping off the plane in New Zealand or the Falklands to see birds, vegetation in all shades of green, and other colours and hues rare in the eternal whiteness of the White Continent. The air, with its smell of vegetation, buildings, and a thousand other scents, was like an olfactory syrup. If one can experience this having come from a place whose temperatures are benign compared to Mars and in which the atmosphere was at least breathable, how might this be intensified by having spent months, maybe years, maybe one's whole life, in pressurised habitats, rovers, and spacesuits in a land with no indigenous life at all?

Herein lies the 'problem' of Earth. Will it not act as a beacon of human liberty to all those people ensconced in pressurised spaces across the Solar System? Will it not call like a Siren to those who dream about its life-giving biosphere? And is it not paradoxical that although many people think of space as an escape from entrenched views on Earth and the rigid tramlines of its history, Earth might soon seem to be an escape from the unyielding lethality of the space environment?

It is true to say that Earth is not perfect. The atmosphere creates storms more powerful than most places in the Solar System (on Mars, the global dust storms are legendary, but the thin atmosphere defangs their ferocity and renders them relatively harmless compared to the worst storms on Earth), and of course, no other planetary body suffers destruction and loss of life on account of a hydrosphere, in the form of tsunamis and floods.

I would not suggest that a solution to this possibility is to launch propaganda about Earth and space designed to ensure the success of space settlements. A free education suggests that people should learn about Earth and, for those who have left it, remember about Earth, in an objective way. With access to the facts, they can make up their own minds. It is likely to be the case that some people's view of their home beyond Earth will not be disturbed. Many polar explorers and scientists feel a strong pull back to Antarctica. I speak to the intensity of this experience myself, despite the evident extremities and life-threatening conditions of that continent. It is hopelessly impossible to explain or rationalise, but it is the other-worldly distance of that place, the shared experiences with others, and the physical beauty in the austere, unforgiving landscape that mix together into a heady cocktail that calls with a shrill, insistent force. There is no doubt that exactly these emotions will be elicited by the Moon, Mars, and every other location beyond Earth. Mars has its own beauty.[15] It is not clear that a long time spent in space would fuel migration back to Earth. With its unpredictable and intricate biosphere, Earth may even terrify some who feel an attachment to the relative simplicity of the lunar and Martian environments. Space will filter and draw certain types of people into it who will have their own reasons to find this a preferable place to live.

Yet I think there is a more fundamental principle rooted in liberty. If it is the case that we must use lies, subterfuge, and propaganda to convince people to stay in space and relinquish Earth, is that a future worthy of humans? Who would commend a civilisation of humans spread out across the Solar System which is kept in place by an intricate edifice of smoke and mirrors designed to dispel an interest in Earth?

It is entirely possible that the reality of Earth could forestall our most capacious ambitions to spread into space, in which case so be it. The matter of Earth should be an open and truthful discussion, wherever humans may live in our Solar System; we must accept with open arms the effect that has on those societies. Perhaps the beauty of Earth and its apparent oxygen-rich freedoms will act as a healthy incentive for extraterrestrial societies to maximise their own liberties so as to hold on to their populations. Earth will silently encourage social arrangements sufficiently felicitous to attract people to stay and to bring new people in. Perhaps, then, the physical liberties of Earth will act as a powerful motive for the social, political, and economic liberties beyond it.

9

JUSTICE AND CRIMINALITY IN THE FREE SOCIETY

Assembling an extraterrestrial justice system

I now want to turn to a matter that intersects with many of the different facets of social organisation I have discussed; this concerns justice and the system of justice in the extraterrestrial environment. As with terrestrial society, we can expect that in any extraterrestrial location, disputes will emerge and individuals will transgress the rules, regulations, or laws that set the boundary conditions for society. From this situation emerges the need for a juridical system to interpret laws, resolve disputes using those laws, and work with the executive to serve justice. The latter includes the need to oversee the physical infrastructure required to implement that justice,[1] which may include buildings dedicated to overseeing those people who have broken laws to a sufficient level of seriousness to merit their temporary or permanent removal from society, i.e. prisons.

The construction of a justice system in this way is a serious undertaking, and it merits its own analysis. For all the reasons that I have discussed about the potential for the arbitrary and coercive treatment of people in the extraterrestrial environment, we can apprehend at once the effort we should dedicate to the means of justice. This focus is not merely about the justice system itself, but the risk of a repressive justice system that is likely to propagate its evils in the social system at large as those who are involved in the system, and the public around them, become inured and succumb to its malignancy.

Fukuyama pointed out that democratic elections are relatively easy to organise, but the creation of a functioning justice system, involving courts, lawyers, an impartial police force, prisons, etc., is a complex challenge to human organisation.[2] His uncontroversial observation emphasises that building a free society involves much more than having a democratic system. The social and physical infrastructure, including that of justice, to make real the decisions of people through the democratic process, must also be assembled. We must attend to that in the extraterrestrial environment from the outset.

Laws must be pragmatic and flexible, not dogmatic

The possibility of errant behaviour or social disorder becoming an existential threat to a settlement would seem to encourage, as in the political and economic sphere, a design for impregnable and predictable laws. Laws would seem to provide the girders to structure society such that personal and collective conduct can be channelled and laid out in ways that can be designed with precision.

An obsession with the objectivity of laws was a characteristic of the French Revolution.[3] It seems that laws were perceived to offer predictability in chaos and to be the essential instrument that could be used to remove injustice. If laws were applicable to everyone and the crimes and punishments that resulted from them were exactly and diligently prescribed by the people, then the chances of capricious judiciaries and decisions motivated by personal and fleeting social circumstances could be avoided. A rational society would thus emerge. At least, that is how the revolutionaries saw it. As Rosanvallon observed,[4]

> The men of 1789 never tired of celebrating the law . . . the celebration of the law derived in part from recognition of the commanding role of the general will, of which the law was seen as an emanation . . . the law was seen as the 'guarantor of liberty' to use an expression that cropped up in any number of speeches to the National Assembly . . . In other words, despotic power was private or particular power (the arbitrary will of the prince), whereas general rules guaranteed liberty.

In the extraterrestrial environment, where there is a strong urge to create predictability against the possible disaster brought on by disorder, and where an entirely new society is trying to establish itself from a collection of individuals who have essentially been transplanted on a tabula rasa, then we could expect that, as with the French Revolution, there would be a strong temptation to regard laws as the apex of social order to which all else must be subordinate.

Laws and their implementation should not be unjust. However, the extraterrestrial society would do well not to attempt to create unyielding legal structures, particularly given that adaptation to new environments may require modification of social conditions as lessons are learned. What may be a serious transgression at one point in time may demand a lesser punishment at another, or vice versa, as the society equilibrates itself to its circumstances.

There are two ways in which an extraterrestrial settlement may avoid this dangerous form of legislative dogmatism and inflexibility. One would be to adopt a common-law approach, whereby laws are established by precedent, and as an extraterrestrial society develops and new forms of transgressions make themselves known, so the pantheon of previous precedents will grow, and judges and juries will converge on responses that reflect the growing experience and knowledge of an extraterrestrial settlement.

The second approach to extraterrestrial law is to adopt the paradigm of civil law, whereby the laws are codified with greater exactness, but society recognises that the relative lack of human experience beyond Earth should encourage a regular review and reconsideration of those laws as experience grows. Carrying out such revisions so regularly that they become arbitrary, under the pretext that the extraterrestrial society is learning, would not be in the interests of society. Nevertheless, every few years, codified laws could be reviewed and revision might occur in circumstances, and at any time, where the law is not fit for purpose, perhaps brought to light by public outrage at specific judgements. Thus, the more formulaic approach of civil law need not obviate adaptability in newly emerging extraterrestrial societies.

The practical implementation of these legal regimes is a matter that needs attention. Quite apart from the theoretical considerations, there are the practical matters. In a small settlement, there cannot be a fully formed court system. Instead, we would expect dispute resolution and the interpretation of the rules and laws to be carried out by a subgroup of the population assigned to this task with the requisite legal knowledge: in other words, space lawyers deployed to such a settlement. Of course, where the settlement is attached to a nation state on Earth, the laws of this originating state may apply, but there is in principle no reason why a small settlement might not find itself, early on, interpreting new local rules, even private laws, specific to that settlement. In larger settlements, comprising hundreds of individuals or more, a traditional form of court structure, with its normalised legal proceedings, could be realistic.

My purpose in this work is not to suggest the details of how a legal structure should be ordered. There is much work for lawyers to do, and there are people with legal expertise and insights who are better placed to debate the exact outlines of an extraterrestrial legal system and its physical manifestation. However, consistent with the focus of this work, my objective is to draw attention to the dangers of assuming that fixed civil laws will provide an elixir for extraterrestrial dangers and that a set of codified laws will bring mathematical precision and order into a lethal environment, enabling the more sure and secure establishment of liberty; this was the mistake made by French revolutionaries.

Laws should be sufficiently predictable to command the respect of those who must abide by them, but they must be sufficiently flexible to avoid them becoming instruments of rigid coercive tyranny. In the long term, a flexibility will allow laws to mutate as a society grows, changes, and learns lessons in the novel environments beyond Earth, making them highly effective at ordering a free society and making the laws themselves one of the features and focuses of continuous democratic deliberation.

The general principles of a justice system

Within these considerations on the judiciary, we might focus on those legal matters that deal not so much with everyday dispute resolution, from private matters to business, but perhaps the most serious matter for an extraterrestrial legal system:

criminal activity. How is criminal activity to be determined, and how are those accused of crimes to be treated prior to their judgement? Unfortunately, the extremities of the extraterrestrial environment encourage a pre-emptive approach to criminality: 'It's better to charge a person with a crime who might be about to damage a habitat and depressurise it early, and ask questions later, than to risk the death of many people.' In essence, as I have explicated elsewhere, as with the social order generally, the justice system has the chance of taking on the flavour of a system working in a public emergency.

If we wish to avoid a collapse into an arbitrary police state, all efforts must be made by juridical, legislative, and cultural mechanisms, mixed with regular overview activities, to prevent the need for safety encouraging a system of arbitrary and heavy-handed justice. We could take this to mean a presumption of: (1) innocence until proven guilty; (2) *habeas corpus*, i.e. the capacity for a detainee or prisoner to go before a court to have the legality of their incarceration determined independently of the people doing the detention; and (3) the prevention of an individual being charged for the same offence twice (double jeopardy). These are at least three cornerstones to act as a guard rail against the tendency for the environment to encourage arbitrary arrest and detention.[5] These mechanisms not only protect the individual from mistreatment but act as effective checks on the power of those running the settlement, establishing boundaries in the effective implementation of justice.[6]

Those arrested for criminal activity must have the right to a fair trial before an independent set of jurors. In the early stages of a settlement, this type of arrangement may be difficult to secure, since the small number of individuals in the settlement will not easily allow for a separate body of jurors that are not prejudiced by their knowledge of the individual concerned, but even under these circumstances, the individual must be given the right to be heard, and it is incumbent on those pressing charges to deliberate the matter as objectively as possible with as much independent oversight as they can. The general point to be recognised is that the isolation and extreme conditions of the extraterrestrial environment give succour to summary and inegalitarian justice. The job of any authority in an extraterrestrial settlement, of any size and at any stage of development, is, in recognising this, to vigorously push in the direction of a culture of transparent and non-arbitrary justice.[7]

Criminal activity is inevitable beyond Earth

There are crimes in the space environment that could potentially threaten many people. They include the deliberate depressurisation of a segment of habitat in which many people live. Although the form of this potential crime is quite specific to the space environment, a similarly motivated person on Earth could aim for mass death by using a semi-automatic weapon. The space environment presents certain types of crimes that are peculiar to that environment, but the outcomes of such deliberate crimes are not necessarily categorically different from crimes with similar magnitudes on Earth. In other words, when we consider the matter of the types of criminal activity we might expect in the extraterrestrial environment, it is not apparent that we should expect an especially different situation and landscape with respect to justice than we have already witnessed on Earth.

A long shopping list of crimes that might be committed in an extraterrestrial settlement is probably not very useful, but we can expect many of them to be identical to those that exist on Earth (for example assault, theft, public disorder, murder), but we can also expect a variety of new types of crimes found only in space to emerge (for example fatal tampering with a spacesuit, damage to life-support infrastructure, attempts to depressurise a habitat, deliberate damage to oxygen-producing infrastructure).

An obvious rejoinder to the preceding discussion is that perhaps, with the application of technology, good selection, and the ideals that may be embodied in a new branch of human society beyond Earth, criminality will be irradicated. We could ask ourselves: is extraterrestrial criminality inevitable? One might think that with everything I have said about art, education, science, and so forth, a well-designed society could avoid this disruption. My objective is to show that no matter how good our intentions, an extraterrestrial settlement will experience criminal activity and the existence of a system of justice follows.

Crude categorisations of criminality are questionable. However, it is useful to make such a separation in order to have a discussion. On Earth, reasons for criminality are sometimes split, crassly, into 'the mad, the bad, and the sad'. Unacademic though this distinction is, it is useful as a first-order position with which to consider the extraterrestrial case.

The 'mad' in a criminal population refers to those people with physiological or psychological conditions that are not evidently linked to social conditions but instead are medical. No matter how good a society we make, we can always assume that there will be births leading to individuals with impaired or altered cognitive functions that may lead to them transgressing the agreed boundaries of society's legal code. Alternatively, such medical conditions may arise later in life. It might be the case that in the future, genetic or pharmacological intervention will be used to help reduce, and even eliminate entirely, the prevalence of genetic psychological conditions that might lead a person to commit certain crimes; this may take the form of the embryonic removal of genes that predispose an individual to high levels of violence or post-birth intervention in diseases that cause cognitive impairment. However, no matter where we think these technological routes may take us or, perhaps more pertinently, the ethical view we have of them, there may always be individuals who, at some point in their life, for medical reasons that were previously undiagnosed or that arise through stochastic physiological or psychological changes, commit a crime. One source of criminal activity in an extraterrestrial settlement is therefore likely to be what would fall under this category. From this initial observation follows the point that crime is likely an inescapable part of an extraterrestrial settlement, even with the grandest utopian visions of removing it. This is a rather trivial, but vital, departure point for recognising that dealing with criminal activity will follow humans to the stars.

The 'bad' refers to individuals who turn to criminality because, for a variety of reasons, they place the criminal activity at a higher value than conformity to society's laws. This decision may come from the individuals themselves, who through observation and deliberate action see an opportunity to advance their ends by breaking the laws, or from peer pressure (perhaps from a gang) to break the law. This branch of criminality can also likely never be eradicated from a society. So long as there exist opportunities for personal advancement through law-breaking, criminality will exist. However, there may be ways to educate an extraterrestrial population to eschew such choices in life, and I have addressed some of these in my discussion on education in the previous chapter.

An example of the emergence of criminality in the 'bad' category would be an individual or a group of persons that seeks to take charge of a vital commodity such as an oxygen-producing machine in a space settlement, in order to control political and economic activity in the settlement. There are ways in which the likelihood of this criminality can be reduced, which I have discussed elsewhere: we might construct social and engineering mechanisms such as decentralisation and modularisation in the settlement to forestall or diminish the ease with which essential products can be seized by those with criminal intent, or we might deliberately mass produce certain commodities in a settlement to reduce single chokepoints in the settlement that could attract criminality. However, no best efforts in social organisation and engineering can entirely prevent such activity.

The final category is the 'sad'. This consists broadly of individuals who are not medically compromised and do not pursue criminality because they see it as a means to an end necessarily, but instead have fallen to the edges of society due to their experience of friends, family, or education. Abusive parents, violent schools, and other factors all contribute to personality types that might resort to criminality either to avenge a perceived wrong committed by society on them, or because it is an easy route to being part of a group, or because their family and educational context have eroded their capacity for empathy and to understand or adhere easily to rules.

This latter category admits of no easy definition, but it is likely to be a significant group with respect to the extraterrestrial society. There are several reasons why we might expect criminality to emerge in space within this category. The extremities within which an extraterrestrial settlement exists mandates a series of rules and laws which finely straddle individualism and the need for a more intense form of collectivism in the injurious environment. This fine demarcation in an environment, where rampant individualism is likely to be regarded as more suspect and dangerous, has the chance of producing certain people for whom this line is difficult to walk; they find themselves unable to divine this balancing act and through their actions, even non-deliberate ones, they end up 'on the wrong side of the law'.

A secondary contribution to the 'sad' category in the space environment is the physical limitation of enclosure mixed with the monotonous external environment, both of which exert on the individual a sense of confinement that is liable to lead to certain individuals feeling detached or alienated from a society which, for whatever reason, they do not feel a part of. This ennui could ultimately lead to criminality if they seek an outlet for the frustration that results from it.

There do exist factors that might attenuate the number of individuals within this category compared to Earth. In the early stages of a settlement at least, the high degree of camaraderie and the efforts expended in creating social cohesion are less likely to lead to criminality from poor educational environments and broken families, phenomena more associated with large urban societies on Earth. However, even in later stages of a settlement, we might expect efforts to be made to minimise these influences. Families that present a risk to the settlement might be moved to the more populous Earth. Once educational establishments of whatever form are constructed in the extraterrestrial environments, we might expect them to be the subject of close inspection and care in a society which will still be small compared to the size of a nation state on Earth.

We could also expect that in any type of settlement, conditions of deprivation would be highly visible to everyone where they did begin to emerge. As a consequence of the lack of a freely breathable atmosphere anywhere else in space, people cannot drift unseen to the 'edges' of society; there are no physical spaces within which people just exist by scavenging for food at the outskirts of a town. An extraterrestrial population is trapped within a pressurised space, and unless we are willing to entertain an image of destitute persons sitting in pressurised walkways begging for water and food from station occupants on the Moon, then all people by default become 'seen' and part of mainstream society: they must be included in calculations for oxygen, water, and food production. In a very real way, these realities work in favour of the impecunious and destitute, even the criminal, since they cannot easily be forgotten. There is more impetus to address serious instances of depredation, and we might expect destitution and criminality to be countered strenuously, except in a very large settlement where anonymity is potentially greater. From the perspective of the justice system, it is less easy for people to exist at the margins of society, unseen by the systems of society that seek to secure order.

Needless to say, the categories of criminality that I have explored are gross simplifications and not mutually exclusive. Sad can turn to bad, and in some cases there may be medical and psychological issues involved. The foregoing is not a sophisticated examination of the reasons for criminal action, but it is sufficiently useful to highlight the general sources of criminal activity that we might find beyond Earth.

The singular point to emerge from the above is that we can speculate ad infinitum on criminality beyond Earth and its form, but we can probably feel confident in making the assertion that no society, no matter how well planned, will avoid it. For that reason, a justice system, and its corresponding infrastructure, must exist in the extraterrestrial environment.

The characteristics of the extraterrestrial police force

As on Earth, laws will be impotent if there is no means to enforce them, and this might lead us to conclude immediately that an extraterrestrial police force will be necessary. In the early stages of a settlement, this may be nothing more than an individual, like a village police officer, being responsible for 'keeping an eye out' for potentially serious infractions and transgressions. As a settlement grows, this responsibility may be taken on by a small group of people, and eventually, if a settlement reaches several thousand people, an organisation that has all the characteristics of a fully fledged police force may take over, similar to one that we might associate with a town on Earth.

Regardless of the size of this law-enforcement function, an indefeasible problem in an extraterrestrial settlement, which I have had occasion to raise before, is the likely open field of play for petty officialdom.[8] Nowhere is the chance of a creeping culture of control likely to emerge more forcefully than in the matter of the implementation of laws and through those charged with the responsibility for preventing law-breaking and the capture of those who do break the law. The reason for this growth in a control culture is the much higher stakes associated with certain crimes, such as threatening the integrity of the infrastructure.

The tendency of those charged with overseeing the implementation of the law to aid and abet a culture of control stems from two sources. First is the human attribute already alluded to: the tendency of some individuals to expand their sphere of coercion into outlets offered to them because of the love of power and importance. There are people who, when given responsibility for implementing a system of laws, will grasp such an opportunity with alacrity, seeing it as a chance to gain some degree of authority, notoriety even, within the community. However, there is a second source of such behaviour, and that is the more benevolent wish to be seen to be doing a good job. No individual would want to be held responsible for having failed to apprehend an individual who deliberately caused a depressurisation event or damaged critical oxygen-producing equipment, resulting in the loss of life, for this would be a clear case of professional negligence. Therefore, the state of mind that this leads to is one of over-caution. The slightest suspicion of activity or behaviour likely to result in criminal activity will cause those charged with law enforcement to act. In the extremities of space, the enforcement of laws becomes keener and more efficient, and those given the job of guaranteeing that they are enforced will move to ensure that this is done. In extreme situations, this may itself lead to police brutality and violence in their efforts to rigidly quell any perceived threat to the settlement; thus, the police force themselves may become one of the threats to the population.

This police force, whether a single individual or a large, coordinated organisation, might be overseen by a public committee to attenuate the tendency towards over-zealous operation.[9] In a small settlement, oversight might be achieved with one or two people responsible for reviewing charges put to individuals by the settlement police officer or officers. In a larger settlement of several thousand, this may be an independent review board, perhaps selected by sortition, who review selected cases of arrest or criminal proceedings to determine their legitimacy. The purpose of independent oversight should not be to make police officers travel in the opposite direction—to become overly wary of pressing their power for fear of being held to account for overstretching their authority. Rather, the independent board should work productively with the police in a mutually improving role to attempt, as far as possible, to keep the power and culture of the police in alignment with the reasonable application of power for the benefit of the safety of the settlement. This oversight is not legal—that is the role of the judiciary; instead, it is to provide the impetus for those on the ground implementing law enforcement to treat individuals in ways that will pass the scrutiny of their fellow citizens; it is the behaviour of law enforcement, rather than the substance of the laws they protect, that could be subject to this observation.

Independent oversight acts as a check on the uncontrolled and unpruned power of a police authority. The effectiveness of this arrangement may be further advanced by publication of instances of law enforcement in the public domain so that the wider settlement may also develop some opinion and view on the appropriateness of the application of police power in the settlement. Annual or regular open public meetings to discuss law enforcement may add yet another avenue to the capacity of the settlers to act as a check and a contributor to ensuring the apposite implementation of laws to the benefit of the safety of everyone, while mollifying the natural tendency of law enforcement to err towards an autocratic frame of mind.

A somewhat unconventional suggestion for an extraterrestrial police force is to charge it with the duty to encourage freedoms:[10] to bring to attention infringements of freedom of conscience and speech rather than to apprehend individuals only for law-breaking that might have caused public disquiet. Alert always to the potential for coercive tyranny in extreme conditions, the apparatus of law enforcement might be actively trained and educated in not merely upholding laws, but also acting to improve the space and opportunity for free expression.[11] Of course, charging a police force with anything other than dealing with transgressions of public order and law may invite the emergence of a culture of busybodies, but the notion of the active encouragement of freedom and individual liberties as a role for a police force is an intriguing one. An example instance would be public abuse directed towards a lone dissenter expressing a point of view in a public space; a police officer might be sent here not to observe the individual, but to encourage them and protect them from crowd passions and to improve their sense of safety as they make their point. In the context of a large protest, one could say that this is no different from police on Earth being sent to ensure that such a gathering passes off peacefully, but there is a subtle difference between their objective being to apprehend individuals engaged in violence and their objective being to actively encourage such a protest and protect the right to free expression. Both of these perspectives can be inculcated in a police force. Here again, an education in liberty and dissent, not just in the police force but in the public at large, can provide the formal framework for grasping these ideas.

The aim of the system of justice

Apart from the obvious statement that a justice system seeks to recognise infractions of laws laid down by a legislature and discourage similar occurrences, we might ask what the fundamental ideals could be of a system of justice beyond Earth: what should be the end goals of such a system? This will influence the types of punishments handed out and the manner in which the physical infrastructure is organised.[12]

Let us consider the aims of the justice system with respect to those subjected to its decisions. In the extremes of the space frontier, a body of criminals 'serving time' in prisons will place a high burden on the settlement in terms of administrative and resource costs. Even those ordered to do community service, perhaps being electronically tagged from their habitats, although not taking up space in a prison, will still need monitoring. It is a simple matter to point out that organising and maintaining a criminal population, particularly one that is physically incarcerated, is a costly enterprise and a burden on a society. However, the costs accrue from more than merely the economic and practical realms. In an extreme environment where individuals are close together in habitats, pressurised walkways, and other artifices, a criminal population can enervate the society, since within it is a sense that society has somehow failed or that it contains a fatal weakness. These are sentiments that are less likely to be felt on Earth, where members of the prison population are often physically quite separated from the rest of society. The proximity of prisoners in the space frontier to an extraterrestrial population will bring forward the ever-present reminder of criminal actions. One solution to this—to remove these individuals to a place where they are distantly separated from the population—may exacerbate the chances that they are ill-treated or forgotten, a point I return to in the next section.

Not insignificant is the fact that a criminal population in any society reduces the number of free and productive individuals playing a full role in the success of the society, and this matter is magnified in a small settlement, where every physical body working for the building up of such a society is of relatively great consequence. If there is little redundancy in available trained people for certain critical assignments, then the forcible removal of an individual from their professional tasks over perceived misconduct may have significantly detrimental consequences on the functioning, and even the safety, of the society. It is, in unembellished terms, a waste of the human resource of a society, and possibly a danger, to have a population of incarcerated people.

From the above, it follows that an extraterrestrial society should have a social system that seeks to reduce the chances of people turning to criminality, an objective that can be met by felicitous social and educational arrangements explicated elsewhere, but also a justice system that seeks to return individuals as quickly as possible to productive and full contributory roles, the achievement of which is largely influenced by the way in which they are treated.

Yet another objective of the system must be to reduce the rates of reoffending. It is a burden on society and its resources when those who leave the justice system harbour a deep resentment against a settlement and its authorities because of poor treatment, and then seek to return to the behaviours that undermine the society they perceive to have wronged them. In a settlement of a few hundred to several thousand individuals, just a few such individuals could cause considerable trouble in the social system, particularly if there is a constant fear that their designs should be directed towards vital infrastructure.

The conditions on the space frontier might therefore lead us to a conclusion that a system of justice should be focused strongly on the rehabilitative function, i.e. the encouragement and restoration of social responsibilities and a sense of capable self-governance to those who commit crimes, while in the process realising the need to enforce a punishment as a reflection of the transgression that occurred. We might frame this in another way by saying that restitution or restorative justice, rather than reprisals, should be the intention of the wider justice system, the latter tending to unhealthy and potentially dangerous outcomes in an already lethal environment.

In cases where the individual responsible for the transgression does not pose a clear danger to the settlement, this mode of justice might recommend itself to community service. Work on cleaning life-support systems, caring for habitats, servicing spacesuits, and so forth, for a certain number of hours, would give the transgressor the chance to atone for their infraction while honing their appreciation for social cohesion and responsibility. At the same time, perhaps they will gain the opportunity to learn a deeper sense of social integration by working with individuals previously unknown or lesser known to them. A community service model of justice may be particularly applicable in space, where the individual, rather than being removed from the society, is instead instructed in productive work of direct benefit to the settlement, while being observed by the community to be seeing through a punishment.

For serious crimes that merit removal from a society, temporarily or permanently, the age-old sanctions by isolated communities such as ostracism or banishment[13] are clearly not practical in the immediate local environment, unless a penal colony has been established in a distant region to which the offender can be sent, but I regard this as undesirable for reasons I will set out in the next section. A potential solution is to remove the offender to Earth, where the much larger population might absorb such individuals and thus remove the burden from the extraterrestrial settlement. In the initial stages of a settlement construction, when the numbers of people are low, this may be a practicable solution, as it is in isolated locations on Earth, such as Antarctic stations. We would presume that at any stage of a settlement's existence, the capacity for physical movement of people to and from the settlement is possible. For the foreseeable future, we can imagine that Earth will be one location with direct links to any nascent space settlement. However, at some point, a permanent settlement will require a penitentiary of sorts, not only to hold people overnight, as in the manner of a small jail in a county police station on Earth, but for longer periods. It is to these facilities that I now turn.

The design philosophy for extraterrestrial prisons

As on Earth, I think that the state of a society is reflected in how it treats its prison population. It is the test of a society's humanity. I do not think it is an exaggeration to say that I also happen to think it is the ultimate test of an advanced social order. A society in which tranquillity and mutual respect are the bedrock of human conduct and where the society has achieved sufficient technical and social development to provide the basic needs to all people to allow for widespread contentment is also a society that finds the time and resource to ensure that even those who have transgressed the laws can be treated with dignity. Prisons may generally be out of sight and out of mind for most of the human population, but look within their cells and listen to the conversation at their lunch tables and the chatter in the exercise yard, and there you will find the measure of the decency of the whole society to which they belong.

It might seem to be of no particular interest to discuss the structure of a prison in space, since we could think that such an institution would be designed like any such facility on Earth, and there is much literature about these buildings and their design.[14] However, there are characteristics of the extraterrestrial environment that should alert us to specific problems associated with the space frontier. More generally, confinement in prison is to deprive an individual of their freedom,[15] and so at once the structure of physical institutions designed to do precisely this with intent must attract the attention of any individual who cares for the conditions and manifestations of freedom in extraterrestrial society at large.

The reader might point out, reasonably, that the extraterrestrial population, enclosed as it will be within habitats and pressurised spaces, is essentially already imprisoned. This could take us on an interesting, and I think eminently worthwhile, philosophical digression on the extent to which humanity is engaging in self-imprisonment beyond Earth. It is true to say that the mainstream architecture of an extraterrestrial society is likely to be much more prison-like than many societies on Earth, and indeed much of this book is motivated by considering the social consequences of that very fact. A fundamental question to arise from this observation could be whether a prison has any meaning beyond Earth: surely it is just another confined space? Prisons do not exist merely to cause confinement per se. They also exist to remove from society people who may pose a temporary or permanent danger and to enforce a curtailment of the liberty of the offender as a form of punishment. The latter objective is one which engenders a great deal of controversy and discussion on Earth,[16] and I have already made the case that in the extraterrestrial environment, where human resources may be limited, a rehabilitative function achieved by involving the transgressor in social tasks is preferable to their complete removal from society. Nevertheless, we might still expect that some sort of incarceration facility may be necessary where the individual is a danger to others. These institutions, as on Earth, are likely to involve more restrictions on movement and behaviour than the rest of an extraterrestrial settlement. So although we could argue that all extraterrestrial settlements are generally more prison-like than most buildings on Earth, extraterrestrial prisons are likely to have distinctive institutional arrangements.

As a prison necessarily involves separating an individual from the rest of the population, it immediately brings forward the possibility of the mistreatment of the incarcerated when they do not receive adequate monitoring by the rest of society. Their relative powerlessness, given their physical situation, and the reasons for their incarceration in the first place, provide a rich and fertile ground for their mishandling. Prison abuse is not a matter I wish to review here, although its manifestation in a more academic environment was given particular force by the infamous Stanford Prison Experiment.[17] However, I do wish to discuss how it may come about.

In the extraterrestrial environment, we have several potentially incendiary factors at play. First and foremost, the incarcerated find themselves in an environment that is instantly harmful, and the threat to use that lethality as a debilitating or even fatal form of force against them immediately springs to mind. They might be threatened with oxygen deprivation.[18] The external environment could become a weapon of coercive control over offenders in a more malign and overt way than how it might be used more subtly against the general population, which I have already discussed in this work. Second, the carrying out of a crime in an extraterrestrial environment that may have threatened potentially large segments of the settlement will bring with it a particularly severe type of opprobrium from the community, giving rise to passions and desires for revenge against the individual who seemed to have so selfishly endangered the lives of many.

It becomes of the utmost importance to prevent both situations from being realised. Recognising the potential use of the extraterrestrial environment as a form of sanction or threat against criminals suggests that a prison should be designed to obviate to the greatest extent possible the capacity to physically use depressurisation, or the manipulation of oxygen and atmospheric content in any other way, as a means to terrorise the incarcerated. This objective could be met by an engineering design that disallowed those running such a facility to depressurise it or parts of it deliberately. This might include walls that are thick and have sufficient integrity to prevent depressurisation; oxygen-provision infrastructure and other plumbing systems that are robustly engineered into the prison and do not have controls that can be accessed by those overseeing the inmates; oversight by independent monitors outside the prison system; other monitoring and recording systems that are accessible by an outside committee in real time, independent of those running the prison.

The management culture and ethos of an extraterrestrial prison system should explicitly recognise the special potential for brutality in the extraterrestrial environment and might even state this formally in its employment contracts, statement of purpose, and design statements for prisons. A draft statement in relation to prison design might take the following form, for instance:

> The deliberate use of depressurisation, oxygen availability, or atmospheric manipulation of any kind to control, incapacitate, or kill prisoners in the extraterrestrial environment constitutes an act of unacceptable cruelty and tyranny. No extraterrestrial prison design shall include the deliberately engineered capacity to depressurise or detrimentally manipulate the atmosphere in cells or areas occupied by prisoners. All design approaches must attempt to minimise the chances that depressurisation, reduction of oxygen, or other forms of control over vital commodities can be implemented within a facility under the pretext of accident or engineering failure.

Once we have a prison population, how are we to realise the objective of returning inmates into the settlement successfully? A prison should be reflective of the environment of punishment, but it should also be a pleasant environment that encourages the strengthening of humanity in its inmates so that they do not become debilitated and brutalised, but rather use the opportunity to improve their education and learning.

We might lay out the basic contours of an extraterrestrial prison. First, we could, in unornamented language, state what we expect a prison to do. I have suggested that an extraterrestrial prison:[19]

> Is a facility to remove criminal danger from space settlements, where criminal damage to infrastructure could kill everyone. Its objective is to harness the productive power of the inmates to the maximum extent possible under respectful and humane living conditions.

A prison should contain all the amenities to give the incarcerated a decent self-respecting standard of living. In Figure 9.1, I have shown the concept for an unelaborate prison. It contains 11 cells, which, crudely using the United Kingdom prison population (0.088 per cent of the total population[20]) would be appropriate for an extraterrestrial population of about 10,000, assuming similar rates of incarceration. The diagram is used to indicate the basic amenities available in a hemispherical-type pressurised dome on the lunar or Martian surface, but of course this depiction could be reduced or increased in area and scope to suit any population size.

The objective is to hone prisoners in the art of self-governance, respect for others, and a respect for the institutions of law: the basic facets of a free society. To achieve this, a baseline requirement is that they have a safe and clean place to rest and sleep, access to clean water and nutritious and healthy food, and the opportunity to learn and advance their awareness of society and the skills they need to survive within it.

The provision of such amenities can never be assurance that the inmates will be treated well. There is always the chance, as on Earth, that the prison population, however small it may be, becomes a forgotten population at the edges of society, figuratively and potentially literally, in physical terms.

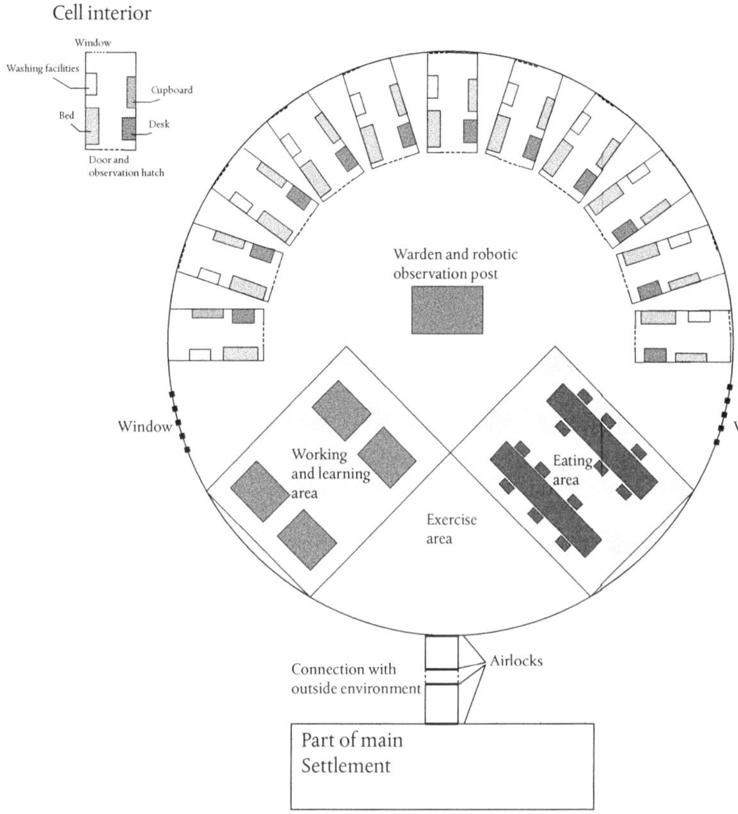

Figure 9.1 A simple extraterrestrial prison for a large population on the Moon or Mars (adapted from Cockell, 2016c). The structure is shown as a two-dimensional facility, consistent with the partial gravity to be found on other planetary bodies. In the open expanses of space, we might imagine this to be rendered in three dimensions. The cell would be about four metres in length.

An antidote to this rejection from society is to ensure that prisons are visited regularly by members of the wider settlement community and overseen by that community. An example of such a mechanism could be that members of the settlement are selected by lot twice a year, in the manner of a jury in some states on Earth, to visit a prison; two to six individuals might be selected at each call. The job of this group would be to visit the prison, speak to wardens, prisoners, and other personnel, and inspect the conditions of the inmates.

The selected participants could elect to travel to the prison at any time of their choosing without forewarning, and upon presenting themselves at the facility, they would be given full access to the institution and its occupants and staff. Through such organisation, requiring no pre-warning, the prison managers would not have the opportunity to create cosmetic changes for the purposes of the visit. The random nature of the oversight visit means that the prison conditions must always be in a state that would meet with the approbation of the visitors. Following the visit, the selected group would have to publish a short report of their impressions and views. This report would be made public.

The purpose of such a system of visits would be not only to encourage prisons to be kept in good condition and its occupants to be treated well, but also to foster in the wider population an interest and concern for the prison population. Selection by lot means that any individual can expect at some point in their lives to have to carry out, and report on, a prison visit. This duty will encourage an interest in prisons, and after having carried out the duty, it is likely that those individuals will be keen to discuss their visit with friends and family, spreading general discussion of prisons and their conditions into the wider polity. In essence, such a system has the dual objective of ensuring that prisons are maintained in good order, inmates are treated well, and the wider population is drawn into involvement and discussion about the prison system and the system of justice more widely.

On the matter of the vocations of inmates in prisons, in the extreme conditions of space, every talent not used is effort wasted. The incarcerated can be encouraged to be involved in productive efforts of benefit for survival and success in the extreme environments of space. Among these we might list servicing of rovers or parts of spacesuits, or assisting in bioengineering tasks for production of fuels and pharmaceuticals. For inmates who have a record of dangerous transgressions involving destruction of public infrastructure, it is probably the case that they should not be involved in matters on which the survival of people depends, such as constructing or servicing crucial components for spacesuits or life-support machines and working on parts of manufacturing systems required for oxygen, water, and food production.

There is a more profound way in which prison education might be developed in space and through which prison inmates could find a highly productive outlet for their creative capacities in contributing to the development of society: they might use their experience of confinement to take an active role in designing settlements and to physically advance the human presence in space.

At the University of Edinburgh, we have run a project called 'Life Beyond' in Scottish prisons for several years, which engages the prison population in designing stations for the Moon, Mars, and other locations.[21] The participants take part in the engineering designs of extraterrestrial stations, but the activity lends itself to artwork, writing competitions (such as writing their first email to Earth), the production of music, and detailed considerations of civic responsibilities and governance structures. The design of societies within extraterrestrial settlements is an open-ended activity that reaches into science, engineering, art, and human organisation, carrying with it the opportunity to advance writing, numeracy, and social skills. In Scotland, this activity was received with enormous enthusiasm and resulted in the publication of two books by the prisoners.[22] Why did this activity work so well?

We believe the reason is that it engaged human minds hungry for the exercise of imagination and offered an opportunity to contribute to constructing an optimistic future for humanity. The chance for a wide range of people with differing interests to engage as a group in this activity made it a challenging but social opportunity.

Not least, prisons are confined environments where contact with the outside world is limited, and there is a high degree of interdependence in the limited internal population. Although there are some dissimilarities, not least that these prisons operate within the Earth's oxygen-rich atmosphere, prisons bear many similarities to planetary stations. We might regard prisoners as having experiences of long-term confinement with direct application to considering social structure, organisation, and infrastructure beyond Earth. The world's prison system is in some respects the longest-running and largest-scale planetary analogue programme yet devised. This may be another factor explaining why participants so successfully, and intuitively, took to this project.

Although in Scotland this activity of getting prisoners to design space settlements is an educational opportunity to use the experience of confinement to plan speculative future societies, we can see that in space itself it represents the nexus of confinement experienced not only by prisoners, but the entire population. In using prison confinement to plan and design further space settlements, the inmates would not merely be engaging in an educational opportunity, but they would be actively advancing the state of society. Thus, prison educational reform becomes inextricably linked to social progress: prisoners themselves can take part in the design of the infrastructure of society as a form of remedial work. This work offers the possibility for deep engagement with the vision of how society is to be developed. Prisoners might use their experience to add authenticity to their ideas, designing future settlements either for the location in which they are living (for example, the Moon or Mars) or speculative settlements for future locations where such infrastructure might be developed (such as Titan or Callisto, or even locations further afield).

Educational opportunities in prisons are likely to be beneficial to the wider society. On Earth, it has been suggested that they reduce rates of reoffending,[23] presumably because they equip inmates with a wider, more appreciative view of their society, providing them with the skills to succeed more effectively after they leave prison. In the extraterrestrial environment, we might consider that advanced educational opportunities would also make inmates more productive once they were liberated.

The philosophy behind all of these ideas is that the justice system should be directed towards the rehabilitation of the offender rather than punishment per se. Of course, a regime that consists of some deprivation of freedom is necessary if prison is not to be an insult to the victims of crimes, but ultimately even they must agree with the general tenet that a society in which we seek to bring those who have committed transgressions as successfully as possible back into the fold is more likely to have peace than one in which those who find themselves on the wrong side of the law are mistreated and end up harbouring a resentfulness, even a vengefulness, against society. This must surely be the case in the isolated, confined conditions of an extraterrestrial settlement.

This view of justice comports with the general thesis of this work, which is to consider how to maximise the conditions for freedom in space against the tyranny-prone influences of the environment. We seek to find a way to bring those who break the law back into the polity and into the cycle of self-governance as productive and content members of a free society. Although this will not always be a success with prison inmates, since some may harbour a deep friction with society or may have committed crimes too heinous to allow for their return, for the majority, this goal is consistent with the wider plans for a social fabric of free government.

10

A FREE COSMOS

When Pericles turned to the assembled crowd and declared to his fellow Athenians that the system of government that they had created was not merely a copy of other arrangements but something to be emulated, an example of good government, he was referring to the freedoms that they had embodied within their culture.[1] 'We are free and tolerant in our private lives, but in public affairs we keep to the law. That is because it commands our deep respect', he said, having just observed that, 'When it is a question of settling private disputes, everyone is equal before the law.'

Despite the simplicity of these ideas and others, such as freedom of expression and conscience, the human character has made it remarkably difficult to implant these precepts into human governance. In some sense, we might express some disappointment that two and a half thousand years after Athens's flourishing of democracy and with the appurtenances of modern communication, education, and good living conditions, more of our fellow humans have not successfully constructed societies that cherish self-governance. This is evidently a struggle that humans must continue. Even in countries where these ideas are thought to be advanced, we might note that not much more than 200 years ago, it was necessary for someone to attempt to vindicate the rights of women.[2]

Now, in the twenty-first century, technological advances have brought to reality the human aspiration to settle the cosmos beyond Earth, to build permanent human settlements on the Moon, Mars, and further afield. As humans expand beyond the home planet, that eternal question of freedom will follow them.

My ambition in this work has not been to make detailed predictions as such; people are notoriously bad at doing that. In addressing the prospects for liberty in space, I have, however, indulged ideas such as the types of education that should be provided and the sorts of art that might be encouraged. However, within all of these speculative ideas, I have attempted to advance two basic claims or hypotheses.

The first hypothesis I have proposed is that although we might eventually build settlements that are so large that the lethal extraterrestrial environment will hardly be on the minds of individuals, the sheer extremity of all space environments compared to the benign environments in which we have evolved and so far constructed our societies on Earth will make space a tyranny-prone environment. 'Prone' is an important word, since I have not claimed that autocracy is inevitable, only that a great many features that space environments share create social, political, and economic conditions friendly to would-be dictators. So my first claim is that freedom will have a harder time in space than it has on Earth. Given the huge cost of tyranny on Earth, in terms of material destruction and loss of human life, we should get to work finding ways to head off this tendency if we intend to spread freedom into space. Perhaps, through these efforts, we may inspirit the human discourse on freedom more generally, and derive new ideas and insights about how to advance it on Earth.

My second hypothesis is that we can use existing methods and develop entirely new tools and approaches to minimise the success of tyranny in space and maximise the chances of creating free worlds across the cosmos. From political to economic considerations, the rebar through the edifice of liberty can be strengthened and designed in such a way as to give it solidity. I have examined four pillars that can be used to hold the building in place; vibrancy in science, engineering, art, and education are essential to the free mind and can in some activities be actively mobilised to enhance the spirit of freedom. I have sought to touch upon dissent and the justice system and how they too might be geared and designed around the precepts of liberty. These deliberations underscore a reality that in various forms has been understood by political philosophers since ancient times: a free society cannot be created by pulling a small number of levers. A constitution, no matter how well devised, will not create a free state. A participatory democracy, on its own, cannot secure an open society. Instead, a constellation of methods and approaches, not least in the culture of a society, must be enabled and sustained for a community to prosper with a free frame of mind (Figure 10.1).

The adversaries of liberty have often led us to believe that it is a fickle and weak thing, that democracy is fragile and easily derailed. They are right. To create the conditions for freedom requires much more than the blunt ignorance and violence of the despot. Freedom is a goal for societies willing to put in the necessary work and with the time to understand its complexities, and those within which individuals are willing to display the forbearance and care to work with their fellow human beings to sculpt and mould the depth of society that allows its flourishing. Just as a beautiful Fabergé egg should not be given to a young child as a toy, free societies that choose the path of autocracy hand to

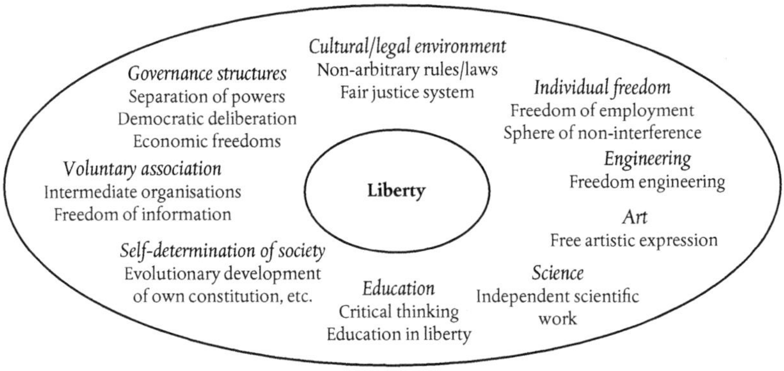

Figure 10.1. A non-exhaustive set of requirements discussed in this work that must orbit liberty for a free society to be stable and successful. All of these must be attended to: a population must nurture and cherish each.

their leaders a thing that will quickly be despoiled. Yet if a population is ready to do that work and put in that time, it will create for itself one of the most astonishing things that has ever arisen from the human mind: the capacity for individuals to live their lives with vivacity and independence of thought, and for society to realise the immense originality, happiness, and variety that emerges from humans whose minds and action are unshackled.

My work here emphasises that space is not necessarily a unique environment in many regards; the political philosophical questions we ask in this environment are not radically different from those that we have been asking for thousands of years on Earth. In many cases it is the intensity of these questions that is magnified by the lethality of space and the influence of its conditions on the societies and forms of tyranny that can emerge. However, we would be right to suggest that certain aspects of the space environment, such as the coerciveness that could surround the provision of oxygen, do bring a uniqueness to some of the challenges. Space is an environment more deadly than any in which humans have dared to create permanent societies, so we would be in error to glibly claim that space is just another location to send people to, a banal extension of our spread across the surface of Earth. The mixture of extremes it poses demands a fresh look at how liberties might be expressed and how we might go about strengthening them.

I do not claim this analysis is exhaustive; I do not suggest that the proposals I have selected are even correct. I have intended to showcase a range of ideas to illustrate that a passivity against authoritarianism, an optimistic blind hope that societies will organically emerge and avoid these outcomes by luck, is unwise. We can plan for liberty.

My view of humanity is not despondent. As an optimist, I hold hope for our capacity to continue to build a magnificent civilisation that is also space-faring. I empathise with Machiavelli and Madison in assuming the worst of humanity with the purpose of achieving the best. By presuming that people will travel the rough road into darkness, we are inclined to consider how to find our way back onto the well-lit path. If we could find that happier path from the beginning, then the torches and maps one prepared to escape darkness could gather dust and we would all be happy for it. Yet, if we should wander down forbidding routes, prior thought and preparation can help ensure a satisfactory conclusion. The purpose of political philosophers of the space frontier is surely to make those preparations.

The exploration and settlement of space offers our species a superlative future. The possibility that in millennia from now, spread out across the galaxy, trillions of our fellow creatures may bring the universe to life with ideas: thousands of Einsteins, Picassos, Newtons, and Van Goghs, generating a scientific and artistic Big Bang. This need not be an unreal dream. For what ultimate purpose? We do not know. Maybe there is none. But if we are to exist, we should try to exist well. Even if these trillions of lives were to be mere moments in a thoughtless and purposeless universe, it is better that those minds live free and joyfully than under the dead hand of oppression, their sparks and glittering effusions snubbed out by the caprice and pettiness of human power.

Whatever the hue of your thoughts about human destinies and purposes, there can be few human beings who do not yearn to see liberty thrive, and where humans go, liberty should follow. On Earth and beyond, our struggle for freedom continues unabated. Let the cosmos be free!

NOTES

Preface

1. De Staël (2008, pp. 753–4).

CHAPTER 1
Liberty on the space frontier

1. Hobbes (2017).
2. Locke (1988).
3. Stated in his preface cover letter to *Leviathan* (Hobbes, 2017).
4. This constant search for a path between the two extremes has been noted by others in modern history. Berlin (2006, p. 195) observed: 'It is a truism that European history is a kind of dialectic between craving for public order and for individual liberty. The quest for order is a kind of fear before the elements, and attempts to build walls and hedges against the chaos caused by absence of control, against the weakening of traditions, habits, rules of life, in an effort to preserve the bannisters that human beings need to prevent them from toppling over into the abyss, to connect them with their past and point a path to the future.'

   Niebuhr (2008, p.108) regarded this problem as universal, as I have assumed it will be for outer space. He said: 'The controversy between those who would "plan" justice and order and those who trust in freedom to establish both is, therefore, an irresolvable one. Every healthy society will live in the tension of that controversy until the end of history; and will prove its health by preventing either side from gaining complete victory.'

   John Stuart Mill (1998, p. 9) also noted: 'But though this proposition is not likely to be contested in general terms, the practical question, where to place the limit—how to make the fitting adjustment between individual independence and social control—is a subject on which nearly everything remains to be done.'

5. Nehru (2010, p. 264). He says before this quote: 'The conflict is between two approaches to the problem of social organization, which are diametrically opposed to each other: The old Hindu conception of the group being the basic unit of organization, and the excessive individualism of the West, emphasizing the individual above the group.' Of course, using the word 'excessive' is question begging, but Nehru recognises the eternal conflict between the collective and the individual in creating a balanced society.
6. The title of Buchanan's 1975 work, *The Limits of Liberty: Between Anarchy and Leviathan*, is an explicit recognition of this balancing act, but almost every book on liberty is at its heart a discussion of this challenge.
7. Dolman (2002); Johnson-Freese (2007); Deudney (2020); Nelson and Block (2018). It might also be noted that science-fiction writers, before the space age and well into it, overtly constructed dilemmas of liberty in space in the fictional context. See Baxter (2015a,b) for a discussion.
8. Nicholson, Schuerger, and Race (2009). The ethical dimensions of planetary protection and space exploration more widely are also explored in the volume by Milligan (2015a).
9. Needless to say, the histories of these two regions do open up very different perspectives on how liberties have been manifested within the different forms of government, religion, political systems, etc. What I mean here is that core questions about political philosophy, such as what sort of freedom of expression humans have, or how governments claim legitimacy, shorn of specific schools of thought, can be asked of any society on Earth.
10. Although there have been many attempts, even on Earth, to link the nature of societies and their liberties to their climatic conditions. Discussed by Bury (2006, p. 79). In this book, my contention is that the extremity of space makes environmental conditions an uncontroversial influence on the nature of societies.
11. Christakis (2019, p. 13).
12. In some sense, I take my lead from Condorcet (1955), who explicated a trajectory of human society and then used it to try to predict the general future of humankind in his 'Tenth Stage'. Here, I try to use

existing political-philosophical literature to predict some general challenges to liberty beyond Earth.

13. We do not know how path-dependent history may be, but like biological evolution, the changes you can make to a society are constrained by what you already have, making it difficult to predict how a society beyond Earth will develop (Moller, 2019, p. 263). A well-known exposition of this type of path-dependent thesis is Weber's (2007) suggestion of a link between Protestantism and the emergence of the culture of Western capitalism. Also proposed by Tawney (1975). An outstanding recent exposition of the emergence of theories of the state and how concepts of freedom were fashioned by religious currents is also to be found in Mortimer (2021).
14. There are a great many anthologies and books that gather together ideas or authors to contemplate the various meanings of the very word freedom. This work cites a selection of them. But a much shortened list across the last 80 years, which includes multiauthor anthologies and which demonstrates adequately the many and varied definitions of freedom, could include: Anshen (1942); Bury (1952); Adler (1958); Muller (1961, 1963, 1966); Oppenheim (1961); Dewey and Gould (1970); Brenkert (1991); Miller (1991); Clark (1994); Popper (1995); Roshwald (2000); William of Ockham (2001); Carter (2004); De Tocqueville (2008); Wilson (2009); Russell (2010); Huemer (2013); Newell (2013); Smith (2013); Siedentop (2014); Cockell (2015a); and many others. For a disquisition on some of the characters that shaped ideas of liberty, see Berlin (1979). A more radical recent view is also Bookchin (2005).
15. Equating to what Priestley (1993, p. 12) referred to as civil liberty, although I shall write about the political means to achieve that, which is what he referred to as political liberty. I am also essentially talking here about 'autonomy', in some ways consistent with a Kantian (2012, p. 51) perspective on human freedom and the autonomy of the will.
16. This culture contains elements of what Berlin (1998, pp. 191–242) popularised as 'negative' and 'positive' liberty. Negative liberty includes those aspects that emphasise non-interference, positive

liberty includes those aspects that emphasise the mechanisms we have (which can include state assistance) to realise our goals and objectives. It should be noted that this 'negative' and 'positive' conception of freedom precedes Berlin (e.g. Bay, 1958, p. 57). Although this is a simplified dichotomy of liberty, as Berlin recognised, I find these definitions broadly useful to capture the non-interfering and proactive parts of human freedom. Of course, positive liberty in particular, insofar as it brings into existence expectations of the capacity to do things, is a contentious form of liberty, a matter which occupied Berlin. However, the expectation of non-interference also makes certain demands on individuals and society.

17. Frankl (2004, p. 134).
18. It would be interesting to explore the sociological and practical implications of a breed of humans that could store oxygen in sacks within their chests, allowing for it to be collected infrequently like the organics they need to burn within it. Some underwater lizards such as anole lizards can collect bubbles of oxygen to extend the time they can spend underwater. However, one could predict that it would hardly influence the extraterrestrial situation, since the consumption of oxygen by the human body is such that sacks of oxygen would not extend the time that a human can be deprived of oxygen for very long.
19. Catling et al. (2005).
20. The extent to which we take for granted the presence of air to breathe as one of our ineluctable and unchallengeable freedoms is to be found in revolutionary literature, in which it is used as the consummate freedom to which other things can be compared. Just one example is this statement by Allen (2020, p. 63), in which he implores his fellow Americans to desist from considering their revolutionary actions rebellion: 'It is no more rebellion for the people to stand up for, and maintain their rights than it is to breathe in the free air.' This comparison disintegrates beyond Earth.
21. Schwieterman et al. (2018).
22. I summarised my view on this as: 'The problem of oxygen is the problem of extraterrestrial economics and, more generally, extraterrestrial liberty' (Cockell, 2009, p. 142). I do believe that the

continuous need for the gas every second of our lives makes it a defining feature of the extraterrestrial environment that takes questions of liberty into realms that we have not generally experienced on Earth.

23. I could instead talk about 'air', but this is an ambiguous term, and the content of 'air' that is breathable by humans can vary. In the US Apollo project, pure oxygen was used as a spacecraft atmosphere. In the Soviet space programme, a mix of oxygen and nitrogen was employed. Scuba divers sometimes use gas mixtures containing helium. All of these mixes can be used under a variety of atmospheric pressures. However, the common factor in all of these mixes is the delivery of the oxygen molecule to the human body; hence, in this work when I talk about a breathable atmosphere, I simply use the word oxygen, accepting that it might be mixed with other components.
24. The triple point is the temperature and pressure at which all three phases of matter co-exist. Below this pressure, liquid cannot exist at all. Mars hovers around its triple point across most of its surface, and this explains the lack of large bodies of persistent liquid water in surface environments (although sustained liquid water might exist in the subsurface). On the Moon and other locations without a substantial atmosphere, the surface is essentially exposed to vacuum. Under these conditions liquid water will boil or freeze, but cannot be sustained in the liquid state when exposed.
25. For example: Ferl et al. (2002); Sychev, Levinskikh, and Podolsky (2008).
26. Hecht et al. (2009).
27. Wamelink et al. (2014).
28. Guerrieri, Lewis, and Matthews (1993); Spudis (2016); Rapp (2019).
29. For example, for carbonaceous meteorites, see Mautner (1997) and for iron-rich meteorites, see Gronstal et al. (2009).
30. Troutman et al. (2003).
31. This work is primarily focused on settlement in our Solar System, but Schwartz explores the ethics applicable to populations on interstellar world ships (Schwartz, 2018). In many ways, his political philosophical conclusions are similar to those we might reach

in our Solar System, but pushed to greater extremes, given the long-term isolation of such a population.

32. I discussed some of these ideas in Cockell (2008, 2010).
33. In his important paper on the emergence of tyranny around water supply, Fuller (1965) describes his own experience of living in a desert and makes this observation: 'One who has had this sort of experience will be left, as I have been, with a lasting scepticism toward the notion that the frontier was a place of "individualism" where every man faced nature by himself. One does not conquer a desert—nor, I suspect, a forest—alone.' This is entirely applicable to the extraterrestrial case. But note that just before, in observing the organisation around the infrequent water supply and the need for control, Fuller said: 'In all this there was nothing remotely suggesting tyranny or autocratic government. Instead there was a strong sense of community such as I have never experienced since. The political issues under most earnest discussion were those affecting the Irrigation District, and everyone had a sense of participating in the affairs of the District. We were all parts, one of another, and we knew it.' Similarly, in this work, I am not claiming that the need for intense control of commodities such as oxygen, water, food, and power will inevitably lead to despotism. It may lead to exactly the opposite social condition. We could not wish for more than Fuller's harmonious sense of community on the space frontier. My objective here is to assume the tyrannical outcome with the intention of maximising the means to prevent it.
34. Fuller (1965) challenged Wittfogel's proposition that irrigation (implying control of the water supply) was always accompanied by tyranny in ancient times because it implied the need for feats of collective coordination. Fuller's point is that invention creates social alternatives to despotism that people will tend to use, hence steering a society away from the inevitability of despotism. In the extraterrestrial case, the tyrannical possibilities inherent in oxygen production may be the very motor for new innovations to produce the gas without despotism. Later in the book, I discuss how engineering may be used to avoid despotic outcomes, but no doubt new social norms and practices will also emerge around oxygen,

water, and food production, just as they have on Earth around water scarcity, as Fuller discusses.

35. I discussed what I mean by tyranny in the extraterrestrial case in Cockell (2010, p. 17). There I concluded: 'In essence it amounts to a gratuitous attack on people's private lives and on their public capacities that serves no purpose other than the advance of power for its own sake, on behalf of individuals or organisations that exert influence over the social order', which is consistent with my general view here, although expressed differently. It bears similarities to a definition of tyranny involving a degree of 'wantonness' discussed by Bryce (1908, p. 335). Often important elements of tyrannous behaviour are arbitrariness and unpredictability, but oppressive and predictable laws can also enslave a population. The common theme between both of these apparently contradictory conditions (arbitrary tyranny and tyranny given shape by a well-defined legal system) is the coercive control of a population so as to severely curtail individual and group freedoms. My definition of tyranny is broadly consistent with one that eighteenth-century revolutionaries came to understand and was adumbrated by Arendt (2016, p. 126): tyranny is a form of government where 'the ruler, even though he ruled according to the laws of the realm, had monopolized for himself the right of action, banished the citizens from the public realm into the privacy of their households, and demanded of them that they mind their own, private business'. However, I stress here that it can include lawless arbitrary power, and the tyrant may not necessarily merely banish people to their private households, because in the worst tyrannies, of course, even this sanctuary is violated— something which Arendt understood thoroughly.

36. In Cockell (2010, p. 16) I discuss a number of ways in which even space autocracies with small populations and few resources might threaten Earth: specifically, control of space resources, threat of dropping weapons on Earth, and the mere psychological effect of staring down into the Earth's gravity well with illiberal intentions. All of these effects and threats may sap the confidence of freedom-seeking people on Earth. I concluded with this view: 'A successful diminution of the conditions that allow extraterrestrial despotism

to take hold may ultimately turn out to be essential for ensuring the continuity of human freedom itself.' I still do not think that this is overdramatic.

CHAPTER 2
The causes of extraterrestrial tyranny

1. See http://www.mars-one.com/ (accessed 15 April 2021).
2. A case for this argument is made by Smith (2016). Zubrin (2011) also sees the Martian frontier as a vivifier of civilisation more widely.
3. Mill (1998, p. 63).
4. See, for example, Milligan (2015b) and Deudney (2020).
5. A point made by Arendt (2006, pp. 260–74) in 'The conquest of space and the stature of man' and discussed by Milligan (2015b).
6. Musk (2017).
7. Schulte et al. (2010).
8. For example: Lewis (1997); Harris (2009); Elvis (2013).
9. It is interesting to note at this point that Montesquieu famously speculated on the role of climate in the way in which laws develop (Montesquieu, 2009, p. 231). In this work, I am not replicating his arguments. Some of his assertions about the way temperature affects people's behaviour might strike us as antiquated and in some parts racist. However, I am claiming that the environmental conditions on other planets, through their effects on, for example, movement and access to oxygen, will influence the social systems that emerge. There is an undeniable strain of Montesquieu in this.
10. I discuss this war-like view of engagement with the extraterrestrial environment and the way in which a society might see itself as mobilised against it in Cockell (2008, p. 259). The treatment of the extraterrestrial environment as a mortal foe to be extirpated leads to all sorts of excuses for centralised directed control, since it amounts to a military operation. As Millar (2006, p. 180) pointed out, 'But in no case is a regulation of this kind so necessary as in the conduct of a military expedition. There is no situation in which a body of men are so apt to run into disorder, as in war; where it is impossible that they should cooperate, and preserve the least regularity, unless they are united under a single person, empowered to

direct their movements, and to superintend and control their various operations.' And again by Russell (1919, p. 146): 'So long as the community is exposed to sudden dangers, the possibility of quick decision is absolutely necessary to self-preservation. The cumbrous machinery of deliberative decisions by the people is impossible in a crisis, and therefore so long as crises are likely to occur, it is impossible to abolish the most autocratic power of governments.' This view applied to the extraterrestrial case is a forceful argument for unity of purpose to the point of intransigence. Arendt (2016, p. 72) also noted the tendency by Rousseau (and his revolutionary followers) to use the common foe to bring wills into alignment (the basis of the general will). The foe here is the physical environment.

11. Nisbet (1962, p. 82) noted the way in which the technical demands of agriculture in the Middle Ages encouraged solidarity among the farmers but equally made them subordinate to the village group. Earlier (p. 50) he noted how these sorts of associations may become more prominent for individuals' lives than other associations, such as the family. Perhaps in the extraterrestrial environments, too, associations forged over oxygen, water, food, and power production and their exigent demands will dominate society.
12. Spiro (1981, p. 54) explains how the extremities of the deserts led to an intense sense of community bond in the kibbutzim. I do not neglect the likelihood that the extremities of the extraterrestrial environment will lead to a similar outcome, but if it does, then there may be little to worry about. My concern here is with societies where the more tyrannical tendencies of humanity are expressed. Spiro states of the early twentieth-century founding kibbutzim: 'These, it must be remembered, were a group of youngsters—"spoiled children," as one chaver put it—who had left their homes and their families and had come to a strange country whose landscape was gaunt and forbidding, whose physical environment was difficult, and whose climate was harsh. They were alone, isolated, and friendless. Within this socio-psychological setting, they were, so to speak, thrown back on their psychological heels; and their strength and consolation—if they were to experience these at all—had to come from within their own group. The psychological

bonds induced by this mutual dependency-security relationship were established in a physical environment of wild but beautiful mountains, and in a social environment free from conventional ties and responsibilities. These factors, combined with their freedom of sexual behaviour and their consequent romantic-erotic attachments, conspired to create what one vattik termed, "a state of hysteria." Given this psychological setting, it is little wonder that they experienced a sense of "organic community." This outcome is possible in an extraterrestrial setting, and those within it may hardly think that they are repressed.

13. Locke (1988, p. 305). Protecting people from physical dangers can always be made to appear as exactly that and not an infringement of liberty. Similarly Mill (1998, p. 107) observed: 'if either a public officer or anyone else saw a person attempting to cross a bridge which had been ascertained to be unsafe . . . they might seize him and turn him back, without any real infringement of his liberty . . . Nevertheless where there is not a certainty . . . no one but the person himself can judge of the sufficiency of the motive.' The complication is that in the extraterrestrial case, the pervasive nature of danger of different kinds leads readily to excuses for these kinds of interventions, where the true danger may be ambiguous, but may look real.
14. The implication of extensive bureaucratic systems for society is explored by von Mises (2007). In an extraterrestrial environment, in order to ensure that no one goes short of vital commodities such as oxygen, and given the possibility of calculating these requirements with exactitude, it would be easy for a sort of Tayloresque philosophy of 'scientific' management to grip the extraterrestrial state (Taylor, 2014) in which 'in the past the man has been first; in the future the system must be first'. There is nothing inherently wrong with applying scientific concepts to planning for items such as oxygen that admit of such scientific accuracy and precision, but when science is used to quash the value of the individual to the point of devaluing human life and opinions, it has been misunderstood, and it is this that the extraterrestrial state must guard against if it is not to descend into despotism.

15. And, of course, if the population itself is completely absorbed in its own survival, it may not require authority to dampen the interest in discussing rights. Rights may not be a high priority for people attempting to survive. Berlin quotes Belinsky, speaking of Russia: 'The people feel the need of potatoes, but none whatever of a constitution' (Berlin, 2008, p. 3). There is a difference here in that many Russian peasants probably were not sufficiently educated to discuss constitutions to make it a matter of priority in the first place. In space, we would expect settlers to be technically proficient and politically aware. However, the point here is that survival can overwhelm the time and intellectual space to get into meaningful prolonged arguments about rights, and this can be exploited by authority.
16. Livy (1982, p. 73).
17. Another fascinating example of this phenomenon is provided by Klosko (2010, p. 175), who describes how the existential threat to the Israeli state made national security a 'superconstitutional premise against which legislation is interpreted'. Appeals to fairness in the interpretation of laws tended to become subordinate to the more immediate needs of security. We can imagine this becoming an almost permanent state of affairs beyond Earth. A just legal system will have to constantly work hard to ensure that fairness is given a place in legal interpretations that may always tend to give the benefit of the doubt to safety in a lethal environment.
18. Just one example is Auguste Blanqui's demands in the face of war with Prussia: 'in the face of the enemy, no more parties, no more differences . . . All opposition, all dissent must give way to the common safety. Henceforth, there is only one enemy, the Prussian', as discussed by Edwards (1971, p. 64). Of course, if you replace Prussian with 'extraterrestrial environment', the point becomes clear.
19. Cockell (2008).
20. Forcefully argued and explored by Michels (1962).
21. This issue of conformity is discussed by Fromm (2006, p. 159). For example, on mechanisms for escape from society relevant to the extraterrestrial case: 'To put it briefly, the individual ceases to be himself; he adopts entirely the kind of personality offered to him

by cultural patterns; and he therefore becomes exactly as all others are and as they expect him to be.'

22. In a way described by Mill (1998, pp. 20–61) in part II of *On Liberty*: Of the liberty of thought and discussion.
23. Alexander (2017, p. 104).
24. Maclean (2006, p. 171).
25. Steel (1965, p. 38).
26. Spiro (1981, p. 99). This statement is a revelatory insight into the extraterrestrial problem. An untenable existence in a kibbutz could eventually lead to an individual returning to mainstream society. However, in an isolated extraterrestrial outpost, leaving may be less easy. Spiro goes on to say that he was told: 'Everyone is concerned about public opinion. Here you can't escape it. Everyone in the kibbutz knows about everyone else. *Sometimes it is too much* [my italics].' Spiro (p. 98) precedes these statements with the simple observation: 'The most effective informal technique of social control in any gemeinschaft is public opinion.' The social challenges, and problems, of the strongly collectivist ethic found in the kibbutz is discussed by Altman (1996, particularly pp. 70–80).
27. Rousseau (1976).
28. For a particular view of Rousseau's vision of liberty, see Berlin (2003).
29. Berlin (1998, pp. 191–242).
30. Although they may be convinced that they are free: 'While the Prisoner of Chillon was not free, it could be said by the poet that he enjoyed liberty, as the "eternal spirit of the chainless mind." Conversely, men can be free to come and go, as the domesticated animals are free, but still be spiritually subordinate to some form of power that provides their material wants' (Morley, 2010, p. 23). The application to the extraterrestrial case is obvious.
31. Frankl (2004, pp. 74–5).
32. These paradoxes are explored in Cockell (2015b).
33. This paradox that Enlightenment values, which sought to free people from arbitrary power, were later themselves perceived as being dull, rational attempts to suppress unshackled individualism by romantics (Berlin, 2003, p. 196) is a warning that too much emphasis

on individualism and a systemic suspicion of order and organisation can itself be highly corrosive to society, and might itself inspire an aggressive authoritarian counterculture. The severity of the extraterrestrial environment requires a balance between the rational order made necessary by science and technology, with the apparatus of survival that emerges and the need for individuals to express their own proclivities. See also Berlin (1999).

34. As Grayling (2009, p. 163) puts it in discussing Isaiah Berlin's view of liberty: 'Even if the critics of Enlightenment are right; even if there is an unresolvable conflict of competing values; even if there is no clear answer to how a given dilemma should be resolved; still, says Berlin, tolerance and reason can help to maintain the equilibrium which is, as he put it, the "first requirement of a decent society".'
35. For example, Haqq-Misra (2019) discusses how 'deep altruism' might sustain space settlement.
36. Of course, this discussion preceded Hayek and von Mises and was a prominent point of deliberation in the Enlightenment for Smith, Ferguson, and others. However, the twentieth century brought debate about its importance in relation to centralised state planning to particular attention, given the large scale of nation states and their capacity to wage war.
37. Jouvenel (1993, p. 130), in his classic analysis of power observed: 'Every authority is, by the law of its nature, essentially dualist. Being ambitious, each separate authority tends to grow; being egoistical, to consult only its own immediate interest; being jealous, to pare down the role of the other authorities. There thus ensues an incessant strife of authorities.' He had much else to say along similar lines.
38. Mill (1998, p. 15). However, we might also note that Mill is not that dogmatic, and his views on liberty do allow for some sort of gradation. For example, in his tract 'Of the limits to the authority of society over the individual' (p. 81), he states: 'The acts of an individual may be hurtful to others, or wanting in due consideration for their welfare, without going the length of violating any of their constituted rights. The offender may then be justly punished by opinion, though not by law.' The point is that there is a gradient

between 'self-preservation' and 'wanting in due consideration' that may be sufficiently circumscribed by laws to protect the individual and prevent a wide range of actions being stymied under the claim that others are acting in self-preservation.

39. Hobbes (2017, p. 275). This view, that safety is the responsibility of rulers or the supreme 'law' precedes Hobbes and was proposed by Cicero in his work *On the Laws* [*De Legibus*] (2008, p. 152: *Salus populi suprema lex esto* or 'To them the safety of the people shall be the highest law').
40. As Hobhouse (2009, p. 35) observed: 'The weaker man consents as one slipping over a precipice might consent to give all his fortune to one who will throw him a rope on no other terms. This is not true consent. True consent is free consent, and the full freedom of consent implies equality on the part of both parties to the bargain.' In the extraterrestrial case, the settlers are thrown the rope of access to oxygen, yet they are under servitude if they are in Hobhouse's metaphorical condition.
41. A sort of extraterrestrial town that has overtures of nineteenth-century industrial towns overseen by a benevolent visionary may not be an unlikely outcome (Kasson, 1999). However, crucially this must not degrade into a town where all the occupants are regarded as lowly workers whom the visionary has 'saved' from a terrible fate.
42. De Tocqueville (2008, p. 123).
43. This is no minor matter, as the notion of liberty as a private sphere free from arbitrary coercion is an important aspect of the very meaning of liberty (e.g. Hobbes, 2017). The history of this emphasis is nicely explored by Patterson (1991, pp. 376–406).
44. For example, in a manner akin to Nozick (1974). Can private exchange be relied upon to produce all the necessary oxygen for a settlement without a central form of authority? Even if it could in theory, surely we must ensure that something required on a second-to-second timescale is not left open to possible market failure. As stated in the main text, when we consider Nozick's work in relation to outer space, we are confronted with a rather essential question: what is the theoretical minimal state in an extraterrestrial environment?

45. Sen (2001, p. 113) notes for the terrestrial case: 'The loss of freedom in the absence of employment choice and in the tyrannical form of work can itself be a major deprivation.'
46. I discuss how religion might become a defining and important bulwark against the extraterrestrial environment and the way in which it might come to dominate a settlement's social organisation in Cockell (2008; p. 267 and 2009, p. 148).
47. Maclean (2006, pp. 126–9).
48. Robinson and White (1986, p. 108) also identify freedom of conscience as an important ingredient in space societies.
49. Asma (2018) provides an intriguing synthesis of the uses and value of religion. The issue I take with some of the arguments for religion is that, however powerful they are, for example providing an antidote to grief, they still do not negate the fundamental question of whether the existence of a God or gods is true or not. We should not believe in untruths, even if they can do us enormous good as individuals, or even stabilise society. Nevertheless, Asma touches on spiritual approaches such as Buddhism, which emphasise a form of internal outlook, control of human desires, and acceptance of our ultimate mortality. These value systems do not require a deity as such, but they are usually included in 'religion'. They would no doubt find a welcoming home beyond Earth in continuously deadly extremities. For some individuals, these forms of religion may even be necessary for their mental health and continuity in such forbidding and lethal conditions.
50. Alexander (2017, p. 105).
51. And cultural convergence has been documented in human societies on Earth. See, for example, the excellent exposition by Christakis (2019). In particular p. 373. The convergence of terrestrial societies faced with similar challenges has also been observed by Fukuyama (2012, p. 52).

CHAPTER 3
Building free societies in the cosmos

1. As Rousseau pointed out, humans are the only creatures that can consciously work to improve themselves (as far as we know).

However, the capacity for perfectability does not imply that we can ever reach perfection, and this is the common erroneous assumption behind utopias. Passmore (2000) provides a wonderful historical perspective.

2. Bacon (2008). This citation also includes More's *Utopia* and Neville's *Isle of Pines*.
3. Harrington's *Oceana* was picked up by Hume (1777), who attempted to emphasise the need to build more perfect commonwealths that did not require imaginary and utopian assumptions about the possible changes to human conduct.
4. Adams (2020, p. 239).
5. Madison (2010, p. 252). As Ostrom (2008, p. 36) observed on Madison's views, 'Since perfect solutions cannot be expected, the practical formulator of constitutions must consider the disadvantages in relation to the advantages inherent in alternatives rather than consider the existence of disadvantages to be evidence that a solution is not perfect, and, therefore, to be rejected.'

It should be recognised that Madison (1788) was not dismissive of humanity's capacity for good. Indeed, he thought it impossible for a government to exist without some inherent virtue: 'Is there no virtue among us? If there be not, we are in a wretched situation. No theoretical checks—no form of government can render us secure. To suppose that any form of government will secure liberty or happiness without any virtue in the people, is a chimerical idea. If there be sufficient virtue and intelligence in the community, it will be exercised in the selection of these men. So that we do not depend on their virtue, or put confidence in our rulers, but in the people who are to choose them.' His point was that alongside this virtue there are inevitably less desirous attributes of humans at play and a government must be organised around these, as much as it attempts to make good use of human virtue. Hamilton (2010, p. 371), writing in *The Federalist No. 76*, had expressed a similar view a couple of months earlier, emphasising that 'universal venality' was as much a fiction as 'universal rectitude': 'The supposition of universal venality in human nature is little less an error in political reasoning than the supposition of universal rectitude. The institution of delegated power implies that there is a portion of virtue and honour among mankind, which may be a reasonable foundation of confidence. And experience justifies the theory: it has been found to exist in the most corrupt periods of the most corrupt governments.'

6. It might be pointed out that this sensitivity was also underpinned by a deep scholarship and understanding of ancient constitutions and their failings among many of the founding fathers. A case in point is John Adams, who examined in detail modern and ancient constitutional arrangements to arrive at his views. See, e.g., Adams (2020, pp. 137–366). There were others who pitched into the fray with their views on the early republic and the need to be vigilant against the inevitability of people with less altruistic intentions. A noteworthy example was Dickinson (2020, Letter XII, p. 95).
7. Otis (2020, p. 46).
8. Macaulay (2018, p. 9).
9. We are reminded by Necker (2020, p. 8) that: 'The barriers that divide the savage from the civilized man, are much stronger in appearance than they are in reality; they were erected many ages ago, and their very antiquity offers itself to our imagination as an index of their indestructability. But it is no less true that a few simple moral principles constitute these barriers, and one or two of these principles, pushed to an extremity, would suffice to unite the spirit of independence and the spirit of tyranny, the equality of the early age with the corruption of the later one.' It is the extremity of the extraterrestrial condition that we should consider as an incubator for challenges to this diaphanous barrier; the primary duty of civil governance in space is to guard against the destruction of this separation.
10. Machiavelli (2008, p. 28). And as Röpke (2008, p. 48) observed in the economic sphere (specifically the conditions in an economy where division of labour occurs, as we can expect in the extraterrestrial case), 'The great risks implicit in an extreme dependence of all individuals in society upon each other are tolerable in the long run only where an efficiently administered legal system and an unwritten but generally accepted code of minimum moral precepts assure

to the participants in the division of labour that they will be able to carry on their activities in an atmosphere of mutual confidence and security.'

11. Mill (1835).
12. Friedman (2017, p. 39).
13. For a recent discussion, see Lukes (2005).
14. Of course, this does not deny that institutions are created to serve people, to serve individuals, as Herbert (1978, p. 249) colourfully explained. Institutions are created by essentially separated individuals banding together to bring them into existence. The matter I wanted to emphasise here is that centres of power are an inevitable feature of society and are being created and exist all the time. Society does not precede institutions as such.
15. The 'state of nature' has had a powerful intellectual influence, but I do not think, from the point of view of the discussion here, that it need necessarily influence how we think about an extraterrestrial governance structure. People living in an environment where a lack of oxygen will kill them instantly clearly are better off if there is some type of organisation in society. In that regard, the notion of a 'state of nature' as a departure point for considering political organisation in any practical sense becomes irrelevant. The question is what sort of social structure is best to minimise despotic rule.
16. Smith (2010).
17. Indeed, the idea that the earliest human societies began with dominant individuals, in essence dominating centres of power, assuming leadership roles, and directing the course of events and the fate of individuals around them, stretches back to Polybius (2010, p. 374), who averred, 'Under these circumstances, it is inevitable that anyone with exceptional physical strength and mental daring will take command and set himself up as ruler over the rest . . . It seems plausible to suggest, then, that this is what primitive human life too was like—that people formed bands, as animals do, with each band following the strongest and most aggressive man among them.' I am not proposing that Polybius' account is the basis of a robust modern analysis of early human society, only that he probably gets at a fundamental truism that centres of power have been a hallmark

of human political organisation from the very beginning of our societies.

18. For example, de Ruggiero (1966, p. 19): 'Ever since the feudal period, and with renewed vigour in the age of absolute monarchy, the Catholic Church has fought against state supremacy; and the very fact of this conflict between two great powers has been an effectual safeguard for individuals against the peril of utter enslavement to either.'
19. This description of power centres in society was developed by de Jouvenel (1993). Although Jouvenel's view is often taken to be antithetical to the liberal emphasis on the individual, I personally think two are completely in congruence. Jouvenel's views that the history of society has primarily been about the interactions of power centres is in some sense a trivial truism. Any form of individualism is necessarily bound up within these centres of power. Indeed, one could say that the whole point of liberal concepts such as the rule of law and freedom of expression is an overt acceptance of Jouvenel's thesis—in other words, the need to protect the individual and to find a domain in which the individual can exist relatively unmolested when surrounded by the inevitability of society's power structures with their immense potential for influence over the individual and their tendency to expand and to use individuals in that process.
20. Cockell (2011).
21. This comparison between modern and ancient liberty and the latter's apparently greater sense of responsibility to society as opposed to the individual was also discussed by Ferguson (2007). It is also eruditely explored by Siedentop (2014), who emphasises the importance of the family in defining early society and the way in which that propagated throughout societal structures, even long after the city had taken centre ground in Greek and Roman societies.
22. Hansen (2006).
23. Let it be stated that this is a simplification of a general view of the relationship between the individual and the polis. It appears to have been taken on by the followers of Protagoras, but the more radical

of the Sophists seem to have held something akin to an extreme individualism. The different Greek views of freedom are explored by Patterson (1991). For just one more modern version of this idea of liberty as participation in the wider community, see Breyer (2008).

24. Reported by Thucydides and cited by Finlay (1985, p. 30), who goes on to examine the concept of democracy in ancient Athens.
25. This view was endorsed by Bosanquet (1920) in his writings.
26. Spiro (1981, p. 28). Spiro also brings to attention other potential contradictions. For example, discussing the philosophy and founding of the kibbutzim, he notes (p. 31): 'The morally sensitive person eschews personal ambition and a desire for a better personal life in order to work for a better world.' Of course, this is somewhat question-begging, since it assumes that one cannot have a personal life while also contributing to a better world. More to the point, by pursuing one's own personal ambitions, one might actually be contributing to a better world, because it is through attentiveness to one's personal life that one might become fulfilled and at one's best, which is to the benefit of society. Nevertheless, these latter points need not nullify the observation that in extremes, a stronger sense of group identity may be an ineluctable part of one's individuality. These same types of observations have been made by Barkai (1972) who emphasised the importance placed on the work ethic in the kibbutz.
27. This psychology explains the seemingly contradictory views expressed by young people who joined the Nazi Party, who conveyed their sense of liberation in committing themselves to its unyielding ideology. It provided a structure to life (Alford, 2005, p. 97). However, when liberty is reduced to a total submission of one's life to a political or economic power, we can hardly condone this or describe it as 'just another form of freedom', for in such a condition lies servility. In a similar vein, Le Bon (2005, p. 43) stated: 'In crowds the foolish, ignorant, and envious persons are freed from the sense of their insignificance and powerlessness, and are possessed instead by the notion of brutal and temporary but immense power.' And: 'The notion of impossibility disappears for the individual in a crowd' (p. 30). Arendt (1976, p. 249) noted this

with respect to totalitarian movements: 'These men began to tell the mob that each of its members could become such a lofty all-important walking embodiment of something ideal if he would only join the movement. Then he no longer had to be loyal or generous or courageous, he would automatically be the very incarnation of Loyalty, Generosity, Courage.' The applications of these pathologies to a small group commanded by a determined individual or group within the lethal conditions of space are quite obvious.

28. De Tocqueville (1998, pp. 71–2). Also observed by Le Bon (2005, p. 82): 'whereas the countries in which the most oppressive despotism is rampant are the Spanish American Republics, in spite of their republican constitutions. The destinies of peoples are determined by their character and not by their government.' This should not be misunderstood as implying that constitutions cannot be used to lessen despotic regimes or that people have no control over the sort of government they create. The point is that constitutions in themselves do not guarantee liberty.
29. Haraszti (1952, p. 221).
30. A similar view is expressed by Jaffa (2004, p. 281), who observed, in a disquisition about Abraham Lincoln's view of liberty, that: 'Only when the purpose and therefore the limits of majority rule are recognized can the decisions of the majority gain acceptance. A free people is not merely one that is under no compulsion by an alien government. It is a people committed in its mind and conscious to the principles of freedom—of minority rights no less than majority rule.' In other words, the commitment to freedom is as much a mental state as it is about formally recognised rules.
31. This is a point also discussed by McClellan (2000, p. 4).
32. Quoting Machiavelli ('just as good morals, if they are to be maintained, have need of the laws, so the laws, if they are to be maintained, have need of good morals'), Pettit (2010, p. 242) explores the importance of norms in establishing the legal basis of a republic. Note that so-called 'norms' can transmogrify into a tyranny of custom, so although we might well agree that an extraterrestrial society, as with a society on Earth, might need norms to instantiate

a stable social order, we also need the means of dissent to challenge those norms, which I discuss in a separate chapter.

33. Inglis (1776, p. 18).
34. Murphy (2018, p. 242).
35. This includes creating an educated populace that can engage in open democratic deliberation. This is nicely explored by Brown (1995) for the case of the American Republic.
36. There are already some examples, such as the Tharsis Montes Commonwealth Constitution and others, reviewed and discussed by Bhatt (2015). Most of these proposed governments are rather specific in their ideas, but I would note that most of them actually conform to the broadly classical liberal ideas that I explicate in this work. I veer away from attempting to design a detailed government in this work; my focus is rather to consider the general challenges presented to liberty.
37. See also Cowley (2019) and Greenall-Sharp et al. (2021).
38. The first amendment states: 'Congress shall make no law respecting an establishment of religion, or prohibiting the free exercise thereof; or abridging the freedom of speech, or of the press; or the right of the people peaceably to assemble, and to petition the Government for a redress of grievances.'
39. The fourth amendment states: 'The right of the people to be secure in their persons, houses, papers, and effects, against unreasonable searches and seizures, shall not be violated, and no Warrants shall issue, but upon probable cause, supported by Oath or affirmation, and particularly describing the place to be searched, and the persons or things to be seized.'
40. Circumscribed by Greenall-Sharp et al. (2021) as 'basic physiological needs', although note the potentially greater number of these in the extraterrestrial environment.
41. Paley (1821, p. 56).
42. I am grateful to Jonathan Lim and Rowena Christensen (2020) for sending me a draft of their document on 'Thriving Not Just Surviving: The Right to a Breathable Atmosphere' produced as part of their *Jus Ad Astra* project. In this document, they discuss the link between the right to a breathable atmosphere and existing legal

frameworks. For example, the right to clean air, although not a formal international 'right', is enshrined within certain national laws and can be extracted from the deeper—e.g. moral—view of the fundamental necessity for humans to breathe in order to realise all of their other capacities for rational agency, etc. They describe how Article III of the Outer Space Treaty, which states that activities in space must be carried out 'in accordance with international law, including the [UN Charter]', provides a link to international human rights obligations (e.g. International Human Rights Law), and thereby to the provision of a healthy environment, including clean air. The relevant point is that although rights to oxygen and air could be constructed from scratch by extraterrestrial settlers, there are certain terrestrial legal and international precedents from which they could come or that would at least provide a useful template for extraterrestrial settlements to debate and devise such rights.

43. Schwartz (2021).
44. Even on Earth, it hardly needs to be stated that there is much contention about what human desires and needs merit the status of formal 'rights' that place demands and obligations on governments. For example, see Shue (2020).
45. We might note some cogent points raised by Shue (2020) that have direct applicability to the extraterrestrial case. He observed: 'A fundamental purpose of acknowledging any basic rights at all is to prevent, or to eliminate, insofar as possible the degree of vulnerability that leaves people at the mercy of others' (p. 30). On the subject of producing sufficient quantities of required items, he stated (p. 46): 'The tragedy of scarcity, beyond the deprivations necessitated by the scarcity itself, is that scarcity tends to make each one's gain someone else's loss' and 'A call for social guarantees for subsistence in situations of scarcity is not a call for intervention in what were formerly private affairs.' Many people of a more libertarian frame of mind may find Shue's arguments unpersuasive, but even dyed-in-the-wool libertarians need to recognise the need for basic levels of subsistence to be guaranteed in an extraterrestrial settlement for the expression of liberty to be meaningful.

46. To follow Paine's example, it is better to deliberate rationally on the structures of government in such extremities than to allow them to rely on time and chance to produce a non-tyrannical outcome (2004, p. 44): 'And when a man seriously reflects on the precariousness of human affairs, he will become convinced, that it is infinitely wiser and safer, to form a constitution of our own in a cool deliberate manner, while we have it in our power, than to trust such an interesting event to time and chance.'
47. Cockell (2009, pp. 140–1).
48. This view of individualism constrained by accepted dependencies aligns with a republican vision of government (Alford, 2005, pp. 38, 104).
49. Of course, a central authority, by provisions of law and a system of justice (the latter existing even in the smallest initial settlement) can also prevent a group of people from descending into barbarism and a Hobbesian war of all against all (Tachibana, 2019). This would be the most disastrous social outcome for a space settlement. I do not discuss it in detail here because if the society should decline into such destruction, then there will be no liberty or tyranny left to consider, although of course, to remedy its impending destruction, the settlement might resort to an authoritarian governance structure to restore order, bringing it back into the purview of my work here.
50. Alford (2005, p. 73).
51. Although terrestrial governments the size of the US government are dissimilar to the types of government foreseen in relatively small space settlements, a ratcheting effect of growth of government (Higgs, 2012) may still apply as their powers grow in response to their growing responsibility to safeguard a population in space as the size of a settlement expands.
52. Stated with characteristic clarity by Thomas Paine (1999, p. 128): 'Government is not a trade which any man or body of men has a right to set up and exercise for his own emolument, but is altogether a trust . . . It has of itself no rights; they are altogether duties.'
53. A point raised in Cato's Letters (1995, p. 405), a set of (sometimes entertaining) letters published by John Trenchard and Thomas

Gordon (*Essays on Liberty, Civil and Religious, and other Important Subjects*). In Letter No. 59 ('Liberty proves to be the inalienable right of all mankind'): 'All governments, under whatsoever form they are administered, ought to be administered for the good of society; when they are otherwise administered they cease to be government, and become usurpation. This being the end of all government, even the most despotick have this limitation to their authority. In this respect, the only difference between the most absolute princes and limited magistrates, is, that in free governments there are checks and restraints appointed and expressed in the constitution itself. In despotick governments, the people submit themselves to the prudence and discretion of the Prince alone.'

54. Macklem (2006) provides a compelling discussion of how these values are the basis of developing independent people in many faculties of their personality, e.g. the role of privacy in gaining the space to develop one's own interpretations and views.
55. The values I elaborate on throughout this work that are consistent with a democratic outlook are also consistent with the benefits of democracy outlined by Dahl (2020), particularly his figure 5 (p. 45) and the discussion that emanates from that list.
56. Of course, the word 'democracy' can, and has been, applied to systems in which people have some channelled involvement in the political system, and this is even possible in totalitarian states (Talmon, 1961). The word is contentious and much misused. In this work, I use it in the sense that people have a genuine choice in how matters are determined within a settlement, where those choices are not highly restricted and pre-filtered by a central authority. The people themselves, when speaking openly and without a sense of coercion, would declare their system to be 'democratic', not merely the authority running it.
57. Arendt (2016, p. 221).
58. The reader should not adduce that by proposing democratic forms of governance in space I naïvely assume that democracy is some panacea for autocracy. It has its own despotisms (Michels, 1962), but I argue that insofar as they attempt to involve people in the process of political deliberation, democratic forms of governance

(whatever their exact form) are likely to be better in the long term for an extraterrestrial society than an unelected system of power.

59. For this suggestion, see McKnight (2015).
60. Recognised long ago and discussed alongside other forms of governance by Aristotle (1988). Despite their age, Aristotle's observations remain as fresh and insightful today as they were when he wrote them. Cicero (2008), in his work *On the Republic*, made similar comparisons on the different forms of government, recommending a mixed form of constitution on account of the weaknesses of more pure forms, including democracy. His views on participatory democracy were largely inspired by the Greek experience (p. 21): 'the supreme power exercised by the Athenian people (to take no other instance) was transformed into the mad and irresponsible caprice of the mob', although his orator, Scipio, later states that 'liberty has no home in any state except a democracy'.
61. A point made by Hayek (2006, p. 91), although he goes on to express the view that it is the best means we have.
62. Finlay (1985). Finlay, however, also explains how demagogues become essential to the functioning of the Athenian democracy. The phenomenon of tyranny in ancient Athens is also comprehensively explored by Teegarden (2014). For a discussion on democracy in Athens, see also Davies (1993). For a similar discussion on the encroachment of tyranny into Rome, see Watts (2018).
63. Necker (2020, p. 269).
64. Burke (1986, p. 151), in a statement that underpins much of his counterpoint to the French Revolution, said: 'Society requires not only that the passions of individuals should be subjected, but that even in the mass and body as well as in the individuals, the inclinations of men should frequently be thwarted, their will controlled, and their passions brought into subjection. This can only be done by a *power out of themselves* and not, in the exercise of its function, subject to that will and to those passions which it is its office to bridle and subdue.' We might agree with the general sentiment that crowd passions can become uncontrolled and should be moderated, but it is not necessarily the case that this moderation must come from an

outside power when it involves political deliberation, otherwise every parliament or democratic institution in the world would require a police force or a monarch to watch over it. This is an overly dim view of human self-control which is not supported by the evidence. Rules of engagement in such institutions are, however, required to maintain order.

65. Polybius (2010, p. 380) observed: 'There were three fundamental building blocks of the Roman constitution—that is, all three of the systems I mentioned above. Each of them was used so equitably and appropriately in the ordering and arrangement of everything that even native Romans were hard put to say for sure whether their constitution was essentially aristocratic, democratic, or monarchic. This is not surprising: the constitution would have appeared monarchic (or kingship), aristocratic, or democratic, depending on whether one focused attention on the powers of the consuls, the powers of the Senate, or the powers of the common people.' Polybius goes on to describe the interactions between these centres of power and the checks that exist between them. Of course, the Roman constitution was more complex than what I am discussing for the initial state of an extraterrestrial settlement, but I merely wish to draw attention to the long-known antidote to the mob rule of a pure democracy: a mixed constitution.
66. In a wide-ranging discussion on government structure which appeared in *Craftsman* in 1734, Bolingbroke (1997, p. 127) opined that: 'Absolute democracy may appear to some, in abstracted speculation, a less deviation from nature than monarchy, and more agreeable to reason, because here it is the will of the whole community, that governs the whole community ... But still it must be confessed, that if it be unsafe for a people to trust too much power to a prince, it is unsafe for them likewise to keep too much power to themselves. Absolute monarchy is tyranny; but absolute democracy is tyranny and anarchy both. From such observations, and many of the same kind and tendency, it has been concluded very reasonably, that the best form of government must be one compounded of these three [monarchy, democracy, and aristocracy], and in which they are all so tempered, that each may produce the

good effects, and be restrained by the counterworkings of the other two, from producing the bad effects that are natural to it.' I think we can have legitimate arguments with Bolingbroke about whether democracies around the world have historically been no less tyrannical than monarchies, but his point about pure democracy tending to tyranny and anarchy is a fair warning made by many others and exhibited in a variety of societies that have chosen that path.

67. Polybius (2010, p. 403). It should probably be pointed out that the Athenian democracy, as represented by the Assembly or ekklesia (ἐκκλησία), was not anarchic. Its agenda was set by the council of 500, whose members themselves were the subject of checks, such as the need to render reports on their activities upon leaving. The point is that the Assembly could be easily swayed by demagogues to take certain decisions and it could become fractious with few checks and balances on impulsive decisions made, as Polybius pointed out.
68. Ostrom (2015).
69. De Tocqueville (1998) was concerned with these conformity-inducing characteristics of democracy. He too observed the power of voluntary association as just one means of encouraging a vibrant public discourse that could help shape a diverse landscape of opinion to lessen the stultifying effects of the tyranny of the majority. The tendency of democracy, in its demands for equality, to destroy the very individualism it purports to encourage was a persistent concern of his.
70. A point discussed by Schumpeter (2008, p. 286).
71. Madison (2010, p. 252).
72. A point made by Skinner (2004, pp. 69–70): 'The thesis on which the neo-roman writers chiefly insist, however, is that it is never necessary to suffer this kind of overt coercion in order to forfeit your civil liberty. You will also be rendered unfree if you merely fall into a condition of political subjection or dependence, thereby leaving yourself open to the danger of being forcibly or coercively deprived by your government of your life, liberty or estates.'
73. The problems inherent in, e.g., Bagehot's (2009) post-hoc justification of monarchy should be self-evident to extraterrestrial settlers.

Bagehot believed that the monarchy was the 'dignified' part of the system and the government the 'efficient' part in a constitutional monarchy. The flaw with this analysis is that these observations are not generalisable insights that can be applied to human societies. Monarchies have not been inherently 'dignified' in their conduct because of some fundamental property, only by happy historical circumstances in some cases. We cannot guarantee this to be the case beyond Earth. His rose-tinted view of monarchies was long preceded by similar but absolutist sentiments expressed by Filmer (1991), who also made a case for a benevolent overseer of society without giving a great deal of thought to what happens when these individuals turn into tyrants with unchecked power.

74. De Lolme (2007).
75. This is an old and hardly controversial observation on one-person rule. Cicero (2008, p. 50), in describing the exile of the Tarquins from Rome, had said: 'You see then, don't you, how a king turned into a despot, and how, by the wickedness of one man, that type of government swung from good to the worst possible.' Sidney (1996) made many similar observations. However, I think that this is a problem for unelected power structures of many kinds, not just the rule of the one.
76. The principle elaborated by Chesterton (1929, p. 35: 'The Drift from Domesticity') that one should find out the reason for the existence of a certain way of doing things before dismantling it, his allegory being a fence across a road.
77. The matter of what defines a society as revolutionary is an interesting question. For example, as Colbourn (1998, p. 3) describes in the opening paragraphs of his magisterial work, in the case of the American revolution, John Adams was emphatic that it was not the revolutionary war that defined the new America as revolutionary, but instead the 'radical change in the principles, opinions, sentiments, and affections of the people'. Here we have a definition that is not about a change in institutions per se, but the attitudes of people that make up a society, which Colbourn examines. I think Adams's observations would likely apply to a new extraterrestrial

population and thus, even from this angle, extraterrestrial society will be revolutionary.

78. It would have been intriguing to know what Edward Coke (2002), the consummate standard bearer of common law, would have recommended in the extraterrestrial condition, where precedent from past laws will be so inchoate. Regardless, his works shine a powerful light on the difference between the terrestrial situation for constructing legal systems about which he wrote, and the extraterrestrial case. His arguments underline how space is a new arena for liberty that is not merely a repetition of these past debates, but the relatively blank canvas of society, and its rapid emergence in locations beyond Earth will throw up very new challenges for the construction of legal systems. One is struck by the potential revolutionary legal nature of extraterrestrial society after reading the meandering and historical contingencies that underpin English common law described in a very readable book by Potter (2015).
79. Arendt (2006, p. 139).
80. The challenge and conflicts in differently sized communities are discussed by Tönnies (2001).
81. For example, as proposed by Price (1991, p. 24), 'From hence it is obvious that civil liberty, in its most perfect degree, can be enjoyed only in small states where every independent agent is capable of giving his suffrage in person, and of being chosen into public offices. When a state becomes so numerous, or when the different parts of it are removed to such distances from one another as to render this impracticable, a diminution of liberty necessarily arises.' Although Price goes on to suggest that the remedy is representative government.
82. Acemoglu and Robinson (2020).
83. Stasavage (2020).
84. Friedman claimed (2017, pp. 126–7) the following for the terrestrial case, which, whatever your view of his particular application on Earth, I think is a serious problem in the extraterrestrial case under extreme conditions: 'In my opinion the fundamental cause of most government intervention is an unholy coalition between

well-meaning people seeking to do good on the one hand and special interests (meaning you and me) on the other taking advantage of those activities for our own purposes. The great movement towards government has not come about as a result of people with evil intentions trying to do evil. The great growth of government has come about because of good people trying to do good.'

85. This comparison between participatory democracy and representative government and its republic forms is discussed by Manin (1997). See also Roberts (1994) for the Athenian case in particular. The ancient Greeks were well aware of the problems of mob rule in participatory democratic systems, Plato sardonically pointing out that you could have so much liberty that even donkeys would walk down the street behaving like citizens (p. 81). I think we would find a concern, even in jest, about self-important donkeys as a criticism of democracy rather strange today, but the point is that democracy does not by default always lead to good outcomes.
86. I first expressed this idea in a discussion with Brian Bergstein, a writer for *The Boston Globe* ('Mars Needs a Constitutional Convention', 5 July 2020) <https://www.bostonglobe.com/2020/07/05/opinion/mars-needs-constitutional-convention/> accessed on 10 January 2021.
87. This is the primary motive for the separation of powers in the first place. The separation of powers in the modern age is attributed to Montesquieu (2009) and then famously to Madison (2010), who justified, similarly to Montesquieu, that through such a separation 'ambition must be made to counteract ambition'. The idea was well instantiated into the Roman Republic and described in some detail by Polybius (2010, p. 384): 'To a considerable extent, then, each of the three components of the Roman constitution can harm or help the other two.' The American founding fathers were well aware of this and cited Polybius in their deliberations about the constitutional structures of republics (see, e.g., McClellan, 2000).
88. McIlwain (2007, p. 132–46) takes a particularly dim view of the separation of powers and its capacity to restrain power, instead insisting on the unwritten constitution and political responsibility as better means of control. I agree with him that there are other essential

characteristics of a political system, such as political responsibility, that are necessary in the extraterrestrial environment as they are on Earth. The separation of powers, on its own, will not guarantee freedoms. However, the separation of powers is surely an effective and essential additional means to place potential checks on arbitrary power and its effects on individuals.

89. Necker's (2020) entire work on executive power was largely motivated by the failings that he perceived in the French Constituent Assembly. Such was the obsession with equality, liberty, and abstract mechanisms to achieve these, that the framers of the new French constitution, he argued, had completely overlooked the need for a nuanced balancing and linking of powers within the government and its branches; 'l'entrelacement des pouvoirs' as he put it—the intertwining of powers—was crucial for the stability of power.
90. I discuss the transition from a centrally planned to a maximally free extraterrestrial society in Cockell (2009, p. 143).
91. Burke (1986).
92. The emergence of the three primary branches is wonderfully explored by Vile (1998, pp. 23–57, and in subsequent chapters). I make the assumption in the text that follows that we wish to separate powers and to dilute power to avoid a centralised sovereign organisation. I accept that absolutists would accuse me of begging the question by making this apparent supposition. However, I hope that, by now, my concerns about centralised authority in the extraterrestrial case are clear enough and the reasons for repudiating the concept of such organisations, such as the popular notion of a supreme 'sovereign parliament' in eighteenth century England—even if such organisations were entirely elected—are obvious.
93. Let me point out at once that historically other arrangements involving the separation of powers have been not only conceived, but implemented. Polybius (2010, Book Six, pp. 380–5) famously describes the separation of the Roman Republic into three branches—that is, the consuls, whom he regarded as being equivalent to the monarchical branch of government, the Senate, which

was an analogue to aristocratic government, and the people, who represented democracy. He detailed the interactions and checks between these three branches and regarded them as essentially bringing into being a mixed form of constitution. Despite the differences with what I discuss, and with what most of us are familiar with, in modern parlance, and making the same comparison as Polybius, we might say that the executive branch is similar to the monarchical branch of government, the executive and judiciary are two aristocratic branches of government, and the people are the democratic branch. We might even say that modern republics therefore have four branches of government if we consider the people to represent one branch.

94. Consider, e.g., Necker's (2020, pp. 32–3) opinion on the unicameral legislature established following the French revolution: 'It is different with a legislative body composed, as in France, of a single house. In that case it soon becomes the object and levelling point of all the passions. Everyone being aware that a majority of opinions is all that is necessary to decide the most important interests of an empire, exterior combinations are formed; individual societies and political clubs put things in train to effect that majority, and assiduously study the art of directing a deliberating assembly; the art of moving it as they please by intrigue, by false reports, by alarming publications, and every other species of influence. All these manoeuvres would be defeated, if the suffrage of two houses was necessary to the formation of laws; the spirit of faction would then cease to be encouraged, and the sources of an infinite number of disorders would no longer exist.'
95. But this is an elected house of fallible human beings. I do not imply that they are philosopher-kings selected to preside over and direct society in the manner of the Republic conceived by Plato (2008).
96. There is no quantitative optimum that has ever been stated or determined for this time; it depends on the nature of the representative body. However, two Earth years is probably a minimum and six to seven years a maximum. Translated to Mars, for instance, two Earth years, rounded to one Mars year electoral cycle, would actually correspond to one Earth year and 322 days. The upper limits on terms

(six to seven Earth years) could be rounded to three or four Mars years, the latter being equivalent to seven Earth years and 193 days.

97. Milligan (2015c).
98. Rawls (1971). This need not involve the fiction of isolated individuals coming together from a 'state of nature', but rather, as Rawls (2008, p. 131) puts it, by 'seeing whether it could have been contracted into in the course of ideal history. It need not actually have been contracted into; a regime may be legitimate even if it has arisen in some other fashion.' Later (p. 172), in examining Hume's critique of Locke, Rawls states: 'His account would be that the present regime is legitimate if it has a form that *could* have been freely contracted into from a position of equal right, even it *actually* reached its present form almost by accident or through various changes over time.' The use of Rawlsian deliberation for considering how to deal with, e.g., orbital debris and space resources, is discussed by Schwartz (2015a).
99. Niebuhr (2008, p. 104).
100. Lucas-Rhimbassen (2021, pp. 31–2) raises the problem of a monopoly established by initial settlers which is then difficult for new entrants to challenge.
101. I have not discussed here the many other benefits of creating a culture of a free flow of information, among them the simple fact that people who read are more informed. These benefits were elaborated for an uncensored press over three centuries ago by Milton (2014).
102. Bailyn (2017, pp. 1–21) provides a remarkable summary of the varied way in which pamphlets were used in pre-revolutionary America and their importance in forming and shaping revolutionary debate. They included short discussions on political positions and rebuttals, literary pieces such as poems, and the development of mature argument. Some of them were powerful and emphatic statements on the right to revolution (for two notable examples of the revolutionary genre of pamphlet, see Mayhew (2020) and similarly Allen's (2021) firebrand arguments on liberty). The pamphlet is a many faceted thing, and although it may have fallen out of favour in modern times, in some sense it lives on in the internet. Perhaps we should expect this style of discussion to be a useful mechanism for

elevating debate in the extraterrestrial case by whatever technical means it can be achieved.

103. A fascinating volume examining the challenges of the emergence of a free press in Eastern Europe is provided by Turfan (1995). The volume is also a reminder of the considerable difficulties that can emerge in nurturing freedom of information when it has been repressed. Once under censorship and control, heroic efforts are needed to reinvigorate an open press. The lesson for the extraterrestrial case is to avoid, at all costs, the repression of freedom of information from the outset. Of course, where this fails, perhaps Turfan's book may be of use to extraterrestrial settlers.
104. Even if one does not agree with Marx, his ideas are worth examining in the context of the conditions in space, and this is done by Cain (2015a).
105. Popularised by Marx (1966, p. 10).
106. Lenin (1932, p. 82).
107. In essence, encouraging a maximally diverse community of human individuals. The importance of human diversity for survival has been pointed out by Robinson and White (1986, p. 103). They were speaking of Earth societies, but within the context of a discussion of building successful space societies.
108. This is not to imply that Sparta was wholly militaristic (Plutarch, 2005). Lycurgus had given the people some measure of say over their laws, but the conformity demanded within Spartan society, and its likely effect on individual intellectual expression that gave rise to the remarkable flourishing of ideas in Athens, is surely undeniable.
109. It is also not clear that a large number of people would voluntarily choose such communes, as discussed by Morris (2002, p. 78). Admittedly, Morris's discussion was about the attraction of arrangements like the kibbutzim in the context of the larger nation state, but given civil arrangements that allow for personal forms of ownership of even just some items, many people would likely prefer this than total communal ownership of all products of human industry. If this is the case, then even a minimal form of private property must follow with the social structures to support that.

110. von Mises (2006, p. 61) makes this point about socialistic economic systems. Here I refer more generally to the social culture, but the points apply to diversity within economic productivity. The importance of social systems that encourage individual liberty and plurality of thought is discussed by Raz (1988).

111. The story of Brook Farm and other socialist utopias is well recounted by Delano (2004). These are many other papers and books that describe these different nineteenth-century socialist communities. One of the most notable is Alcott's personal testimony of the Fruitlands utopian community of the early 1840s (Alcott, 2012). The reasons for failure tend to be similar: primarily a lack of flexibility and a demand for collective behaviour that fails to take into account the realities of individual human proclivities and desires.

112. An account of some of utopian communities is given by Reece (2016). He discusses the longevity of some of them (such as the east coast Shaker communities), but his book also underscores how most utopian communities were usually dependent on the larger society to exist and some of them had quite extreme and cult-like social orders. Indeed, some of them had highly questionable social practices. Nevertheless, many of the effective aspects of these communities, such as decentralised governance and local economic systems, have served some of them well; this is observed in modern self-sufficient communes that are not as religiously motivated or isolated as many were in the nineteenth century.

113. Klosko (2010).

114. Voluntary association has also been suggested as one of the key elements in 'resurrecting' civil society in transition out of authoritarianism (O'Donnell and Schmitter, 1990, p. 53).

115. And this capacity may also be important for finding solutions to new challenges that arise as a settlement develops. Pettit notes on this matter (2010, p. 195) that: 'It is absolutely necessary in a contestatory democracy that people should be able to coalesce around group identities that were previously suppressed, for example, or espouse various causes that were not previously salient, and that they should have the opportunity to bring public opinion and political life around to their point of view.'

116. Nicely explored by Levy (2015).
117. Beneke (2006). The religious history of early American states and its emergence is also explored in fascinating detail for several of the American states in a set of papers by Esbeck and Den Hartog (2019). The trajectories are different for each state and they give some clue as the sheer variety of views and opinions that are possible on religious matters when they are given free play. We might expect highly specific histories, given shape by the extraterrestrial environment, to emerge beyond Earth.
118. Jones (1977). Also discussed in the much older book by Headlam (1890).
119. Discussed by Dunn (1992, p. 60).
120. The way in which the process of lot contributed to mechanisms of accountability in the ancient Athenian direct democracy is nicely described by Borowiak (2011, pp. 79–99).
121. Friedman (2002, p. 23) makes the point that the market 'permits unanimity without conformity'. Of course, this is true, provided markets are not greatly distorted by lack of choice, which could be the case in small extraterrestrial settlements. However, for the extraterrestrial case, encouraging a lack of conformity in an environment where vital resources such as oxygen could fall prey to single producers is the most vital role of a market in different products. Also discussed in Cockell (2009).
122. Radford (1945). Radford's experiences were in some ways a vindication of Menger's historical view (2009) that money would spontaneously emerge without central oversight. The difference being, of course, that in a prisoner-of-war camp cigarettes became the medium of exchange rather than precious metals, as Menger had discussed for the history of money in the wider world.
123. To borrow and modify a famous metaphor: Smith (1991, p. 399).
124. But nevertheless, this 'competition' and self-interest is a form of cooperation: Smith (1991, 1999) and Palmer (2009, p. 172).
125. Haqq-Misra (2016).
126. See Persson (2015).
127. Cockell (2009, p. 144).

128. Arguments that were debated and given a full airing in the years leading up to the American Revolution. See, e.g., Bailyn (2017, pp. 210–29). Extraterrestrial founders and settlers would do well to understand this history, since much thought was given to 'internal' matters (matters within the colonies) and 'external' matters (the proclamations and demands of Parliament), including the matter of taxation, and ultimately to the notion that no middle road of split sovereignty could be found, or was logically possible—a position that ultimately led to the revolutionary rupture.
129. Smith (1991).
130. Hayek makes cogent arguments for the difficulty of central planning and the near impossibility of gathering sufficient information to regulate production of a diverse economy in real time. He also adumbrates the result of a lack of flexibility for the production engineers and operators in a centrally planned system where production is tightly planned and ordered. My point here is that in the extraterrestrial case, even if these problems can be overcome, as might well be the case with the oxygen requirements of a small community, we should wish to avoid this situation because it has a tendency to lead to dangerous political centralisation around something required by people on a second-to-second timescale. This also applies, however, to less continuous requirements such as water and food and many other products linked to the vital necessities of survival in space. We might also note that even in the case of centrally provided oxygen, unpredictabilities could rapidly lead to disaster because deviations from the plan would lead to, as Hayek recognised for terrestrial central planning, 'abandoning the comparatively scientific methods . . . Instead are substituted more and more arbitrary and uncorrelated decisions of particular problems as they are suggested by the contingencies of the day' (1980, pp. 154–5). Similar observations were made by von Mises (1944).
131. Stated succinctly by Hayek (2007, p. 108): 'the system of private property is the most important guarantee of freedom, not only for those who own property, but scarcely less for those who do not. It is only because the means of production is divided among many people acting independently that nobody has complete power over us,

that we as individuals can decide what to do with ourselves.' Hayek's last statement, which implies that we can gain powerful freedom over our lives through private property, may not always be the case, since private corporations can exert considerable control, but nevertheless his point is that a system of private property fragments production, and therefore power, which is plainly of interest for the extraterrestrial case. We might note that if all the private property is assembled under the control of a single individual or group of individuals, then the settlers will have gained little. As Malinowski (1960, p. 251) pointed out in a lucid way of relevance to the extraterrestrial case, 'Insofar as property means an exclusive access to the use of certain instruments or of the ability to deprive others of the use of such instruments or of the benefits derived from such a use, we have potentialities of slavery in the economic sense.' The system of extraterrestrial private property must allow for plurality in ownership and actively encourage it.

132. One does not have to be a card-carrying libertarian to believe in private property in space, but it is simply stated as: 'The right to private property, via property law in developed communities, secures for everyone a sphere of personal jurisdiction so that it can be determined when one is acting in one's own sphere or intrudes on the spheres of others' (Nussbaum, 2005, p. 13).

133. Smith was not the first to draw attention to the way in which self-interest motivates ultimately beneficial public outcomes. One well-known exposition that precedes his is Mandeville (1988). However, Smith made the now famous observation: 'It is not from the benevolence of the butcher, the brewer, or the baker, that we expect our dinner, but from their regard to their own interest. We address ourselves, not to their humanity but to their self-love, and never talk to them of their own necessities but of their advantages' (Smith, 1991, p. 13). After this exhortation, Smith observed: 'Nobody but a beggar chooses to depend chiefly upon the benevolence of his fellow citizens.' I find myself in harmony with Smith's essential observation that humans are self-interested and that a successful economic system should mobilise these behaviours for the wider social benefit through voluntary transaction. However, I also hold

that in the extraterrestrial environment, where oxygen want is both lethal and a fertile offering for tyrants, a virtuous citizenship would have benevolence towards others in the matters of oxygen provision precisely for the purposes of preventing it from falling prey to the despot. This is not necessarily in contradiction to Smith, who shrewdly uses the word 'chiefly' in his statement.

134. The United Nations (UN) Moon Agreement ('Agreement Governing the Activities of States on the Moon and Other Celestial Bodies') adopted in 1979 explicitly states that the Moon and its resources are the 'common heritage of mankind'. It should be noted, though, that this is not stated in the 1967 UN Outer Space Treaty ('Treaty on Principles Governing the Activities of States in the Exploration and Use of Outer Space, including the Moon and Other Celestial Bodies'), which has a larger number of signatories. That treaty states that celestial bodies are 'not subject to national appropriation by claim of sovereignty' which is subtly, but significantly, different, and can be interpreted to allow for local claims on resources and private industrial activities. These matters illustrate that discussions on liberty and resources are already an active area of debate.
135. Rousseau (2004) was phlegmatic in his disdain for private property and its effects on the human character. The problem with applying 'the fruits of the earth belong equally to us all, and the earth itself to nobody' (p. 17) to the extraterrestrial case is that if this is mandated by rules, what motive does anyone have to make the vast efforts to do something productive with extraterrestrial land in extremes that far exceed anything on Earth? Discussions on the relationship of land to space legal regimes and norms can be found in Haqq-Misra (2015) and Rummel (2015).
136. For example, for the Moon, see Elvis et al. (2021).
137. There have been a number of discussions on how to create wilderness areas and parks in space and how to allow for private land ownership alongside these areas. See, e.g., Uhlir and Bishop (1986); Almár (2002); Cockell and Horneck (2004, 2006); Rogers (2004); Willliamson (2006); Bruhns and Haqq-Misra (2016); Matthews and McMahon (2018). Note that these suggestions sit within many decades of planetary protection considerations such as by COSPAR

(Committee on Space Research) that are primarily focused on the protection of scientific value.

138. This is discussed, e.g., in Cowley (2019) and Greenall-Sharp et al. (2021).

139. I am measuring age here in Earth years. It is a very interesting question as to how age is measured beyond Earth. Do people forever use the Earth year as the measure? In this case, birthdays would occur on different days for every birthday, which could be confusing. Do they measure their age in the local calendar? If they did so, then they would have to accept very different social perspectives. A person who would die at aged 70 years on Earth would die just after their 37th birthday on Mars (where the year is 668 Mars days (sols) or 687 Earth days). Although this seems rather depressing for an Earth dweller, of course Martians would become inured to the much longer years of their lives, with a person aged 45 years (just over 84 on Earth) having reached a ripe old age. It may thus also be the case that birthdays will be much more precious as one moves outwards in the Solar System and so come to be celebrated more vigorously.

140. Let me say that I think there is nothing distinctly extraterrestrial about this. A scheme such as this could be effective in local communities on Earth.

141. For example, famously for the Moon, Heinlein (2005). An example of this matter raised in the peer-reviewed literature, a paper which also addresses the likelihood of property right disputes, is Lin (2006).

142. Although with the benefit of hindsight, historical demands for independence seem like defined moments in time, these processes have often been drawn out and built on a slow growth and erratic expression of the desire for separation. The vagaries of the American road to the eventual separation from the British is just one example (expertly explored by Bailyn, 2017). One might say the same of India (Nehru, 2010). In the extraterrestrial case, even if there are desires for sovereignty and independence from Earth, those expectations may linger and grow over protracted periods.

143. Reported in Robinson and White (1986, p. 161). I suspect that Gingrich may have been inspired by the visions of space settlements on these scales proffered by O'Neill (1976).
144. Formally the Treaty on Principles Governing the Activities of States in the Exploration and Use of Outer Space, including the Moon and Other Celestial Bodies.
145. Crawford (2015).
146. This is a matter explored by Smith (2016), who points out that new settlements could generate adaptive diversity which is of value not just to the settlement itself, but to its mother culture. We cannot predict what those new cultures and capacities might be until we actually conduct the experiment.
147. Whitman Cobb (2021) notes more generally the role of commerce beyond Earth in mitigating conflict. I have emphasised the desire for vibrant trade routes across the Solar System to maximise economic liberties, with its knock-on effects on political liberties. Whitman Cobb's analysis shows how, in pursuing such an objective, we might also discourage serious conflict.
148. The word cosmocracy has traditionally been applied to the concept of global government. For example, Keane (2002).
149. Schmidt and Bohacek (2021).
150. The problem of distance and the speed of light in building federations is indirectly considered by Forgan (2011) in the alternative context of contact with extraterrestrial intelligences and their speculative capacity to move throughout the galaxy.

CHAPTER 4
Dissent and welfare

1. The volume by Cockell (2016a) presents a series of essays about extraterrestrial dissent. We might gain some idea of the sorts of dissent that could emerge in an extraterrestrial environment from historical revolt (see also, e.g., Cohn, 2008). But we would expect reasons for dissent to revolve around new concerns such as oxygen or water production, not the papal and monarchic concerns of, e.g., medieval Europe. These extraterrestrial concerns will need

new analysis and new responses. Some reasons for dissent, such as overwork, have already emerged in space stations. Regulatory agreements have been put in place to circumscribe boundaries of jurisdiction. We can expect that such rules would perhaps mutate into instruments more akin to 'laws' in developed settlements. This is explored by Bhatt (2016).

2. Solzhenitsyn (1997).
3. Jefferson (1787).
4. Pettit (2010). This notion of domination, particularly arbitrary domination, as a characteristic of tyranny is also explored by Lovett (2010). Interestingly, earlier John Adams (2014, p. 29) used the words 'domineering spirit' to describe the social effects of the canon and feudal law.
5. Laudani (2013, p. 37) provides an excellent critique and study of the definitions and concepts of disobedience. I do not discuss the theoretical arguments in this work; my focus is on the effects of extraterrestrial conditions on disobedience in a broad sense, caused by disagreement with an acting authority.
6. Zinn (2013, p. 16).
7. As Singer (1973) emphasised.
8. Milligan (2016) explores and defends 'constrained dissent' in the extraterrestrial condition even within the competing needs that individuals have within such settings.
9. Cockell (2016b, p. 26).
10. Cockell (2016b). This assumes that we wish to deliberately create spaces where dissent can occur, and I believe this to be essential in the extraterrestrial case. In the terrestrial case, Delmas (2018, p. 214) states: 'The civic virtue of open-mindedness flourishes only amid social integration. There must be a range of environments favourable to meaningful interaction. When people regularly interact with others who look or act differently from them, they may shed their fears and prejudices about these "others" and stop thinking in terms of "us versus them." This phenomenon has been well documented in sociology and discussed in philosophy.'
11. Zinn (2013).

12. Pettit (2010, p. 56) observes: 'But how to test for what is sectional or factional? The only possible means is by recourse to public discussion in which people may speak for themselves and for the groups to which they belong. Every interest and every idea that guide the action of a state must be open to challenge from every corner of the society; and where there is dissent, then appropriate remedies must be taken.' Of course, in practice this is easier said than done, but the general ethos one should attempt to encourage into society is clear.
13. Bedau (1961, 1970).
14. Gandhi (1961).
15. Cockell (2016b).
16. Milligan (2013).
17. Crouch (1982).
18. Young (1988). But see also the earlier paper by Hopkins (1979) of the RAND Corporation, who examined early strike action in space settlements and came to similar conclusions.
19. Schwartz (2016).
20. In this work, I am primarily concerned with liberty within human settlements. I do not address war between settlements, which is another form of 'revolutionary' activity. The matter of war, and its potentially lethal consequences, is discussed by Crawford and Baxter (2015) and Whitman Cobb (2021).
21. Heinlein (2005).
22. Cain (2016).
23. This interface between terrorism and liberty on Earth is explored by Grayling (2009).
24. This problem was recognised for terrestrial societies long ago and was succinctly summarised by Sallust (2007, p. 14–15): 'For it is always those in a community who have no resources who resent the good and extol the wicked, who hate what is old and crave what is new, who in their hatred of their own circumstances are enthusiastic for everything to be changed, and who thrive on disruption and sedition with no concern, since destitution is a possession easily held without loss.' This is no less a concern on the Moon and Mars than it was in Rome.
25. See also Cain (2015a, pp. 97–8).

26. On Earth, the extent and purpose of the welfare state are a matter of intense debate (e.g. Van Parijs, 1997). So too are they likely to be in the extraterrestrial case. I leave the matter of the detail of an extraterrestrial welfare state to others to debate; I suspect that the conditions would vary depending on different locations in space and the culture and social arrangements that become moulded by different environments. Nevertheless, from the point of view of the basic precepts of liberty, I think most people would agree that given the extremes in space and the desire to reduce indigence and criminality within such confined settings, some type of safety net, welfare state, or whatever one wishes to call it, will be a necessary device.
27. A point raised by Sen (2001, p. 66).
28. This is a matter that is raised by Hogan (2021, p. 26), who suggests the following, referring to terrestrial cosmopolitanism: 'In the same vein, settlements in outer space could become a region to exchange ideas with people of different backgrounds and help promote a more planetary consciousness, as the spirit of solidarity and international cooperation on the ISS [International Space Station] has shown.'
29. The valuable role that people with disabilities can play in space exploration and settlement is discussed by Wells-Jensen, Miele, and Bohney (2019). The authors also discuss architecture, suggesting the importance of using universal design to allow for disability in space conditions.
30. A view of humanity described by Cicero (2003, p. 21)—the fellowship of the entire human race (*universi generis humani societate*), although of course he was not referring to a view of humanity literally realised on cosmic scales.
31. The 'Dream of Scipio', part of Book Six of Cicero's *On the Republic* (2008, pp. 86–94), is an astonishing piece of speculation. Of course, the cosmic order as understood by him was very different from the reality we understand today, but that did not stop Cicero from imagining what it might be like to be taken beyond Earth and to look back. His comments foreshadow the observations made about the smallness of Earth about two thousand years later. Africanus

tells Scipio: 'you can surely see how small is the area over which your glory is so eager to extend. And even those who talk about us now—how long will they continue to do so?'

32. White (1987).
33. This is a matter discussed by Crawford and Baxter (2015). The title of their contribution about the 'lethality' of interplanetary war sums up the essential concern.
34. I might also note that literal slavery is a possible outcome of the extraterrestrial condition. If we were to accept some version of liberalism in space, slavery is impermissible: 'The legal condition of slavery is excluded by liberalism, then, because its defining feature, the possession of one man of property rights in another, is inconsistent with the central, equal-freedom principle of liberalism' (Gray, 1991, p. 81).

CHAPTER 5
The development of science and liberty

1. von Mises (1990, p. 22) is no doubt correct that in a complex market economy, 'In business offices and laboratories the keenest minds are busy fructifying the most complex achievement of scientific research for the production of ever better implements and gadgets for people who have no inkling of the scientific theories that make the fabrication of such things possible.' My view is that in the extraterrestrial environment, where existence is more precariously dependent on such gadgets, every effort should be made to ensure that people do grasp these theories if they are not to be at the mercy of those who wield this knowledge.
2. There is a great deal of debate, and literature, about how science works and the nature of so-called 'scientific revolutions' (e.g. Gower, 1996; Kuhn, 2012; Cowles, 2020). The complexity of the mix of chance, logical thinking, the social environment (the zeitgeist), and elements of genius is explored by Simonton (2004). However, I think that most people, particularly scientists, would have little hesitation in agreeing that chance, made possible by giving at least some people and institutions the opportunity to think about scientific challenges in a free and undirected manner, is an

important ingredient. This will find powerful expression in lethal extraterrestrial environments, where chance should be given its full opportunity to stumble on solutions.

3. For example, see Ferris (2010).
4. For example Neufeld (1999).
5. The nexus of science and space is a topic that has been extensively examined by Schwartz, e.g. Schwartz (2015b, 2020).
6. Qualifications can be a problem because those with a PhD might not necessarily have a commitment to science, having left any involvement in it, and there are of course many outstanding scientists who do not hold professorships in institutions and do not use these appellations, since they are university terms. However, some defined level of productivity, e.g. a certain number of papers on which the individual is a first author and which can be verified as being the original work of that individual, might be a serviceable approach. One should not seek a perfect means, however, since there is none; the point is to be objective.

CHAPTER 6
Engineering liberty

1. See, for example, de Botton (2014) for a wonderful exposition of how architecture affects our happiness.
2. Perhaps one of the better known exemplars of this aspiration is Owen (1991) and in particular the vision he lays out to the denizens of New Lanark (pp. 105–34). I do not aspire in this work to the construction of a utopia, or claim some new vision of how to shape the human character by means of the artificial environment. But I do claim that we can use engineering to thwart the progress of tyranny and to create social environments that allow humans to express forms of liberty in the most effective way possible.
3. Cockell (2019). The planning of liberty, involving engineering, is also discussed in Cockell (2015c).
4. Cockell (2019, p. 3).
5. For example, the way in which habitat design influences the psychological environment in isolated locations is comprehensively explored by Häuplik-Meusburger and Bishop (2021).

6. NASA TechPort (2018); Sridhar and Vaniman (1997).
7. Schmidt, Newman, and Hodgson (2001); Graziosi and Lee (2003); Jordan, Saleh, and Newman (2006).
8. Häuplik-Meusburger and Bishop (2021). Their work emphasises the need for a balance between public spaces that allow for sociable activities, yet the need for private spaces that are also an ineluctable part of human psychological welfare. Thus, any form of freedom engineering must consider not merely the need to devise an architecture that prevents control over resources, but the need within that design to physically lay out a settlement and its interiors in a way that takes into account the balance of public and private spaces required to satisfy human psychological and social wellbeing.
9. See Baxter (2016) for a discussion of this in the context of large world ship-type designs.
10. Sagan (1973); Fogg (1995); McKay, Toon, and Kasting (1991); McKay and Marinova (2001).
11. A principle suggested by Schwartz (2021).
12. Discussed in Cockell (2015c), but elaborated on here.
13. Mason (2021) offers a sweeping vision of the potential for human engineering in space. See also Szocik and Wójtowicz (2019) and Szocik and Braddock (2019).
14. See Cain (2015b) for a discussion on some of the health limits of astronaut freedom.
15. The possibility of 'moral enhancement', for example the introduction of altruistic behaviours to reduce the chances of conflict in space station crews, has been considered by Szocik and Wójtowicz (2019).

Chapter 7
Art and liberty

1. Most extremely expressed for Prometheus, who 'in spite of his horrible outward suffering and indignity [being chained to a rock], his will remains unbroken' (Patterson, 1991, p. 167).
2. For just one example, see Nehru (2010).
3. The Inuit are a fine example, since they occupy the extreme conditions of the north, not dissimilar in some environmental characteristics, such as low temperatures, to Mars (although in most respects

still considerably less extreme than Mars). A beautiful example of their work can be found in Hessel (2006). An illustration of the influence of their environment in a diversity of artistic forms is to be found in Finlay (2000).

4. Cockell (2015d).
5. One of the facets of art lauded by the ancient Greeks as described by Agard (2018, pp. 102–15).
6. The psychological and engineering considerations involved in space habitat design are considered by Häuplik-Meusburger and Bishop (2021). They also discuss the psychological problems of isolation and confinement in space that shape the social environment.
7. Cockell (2002).
8. As Muller (1960, p. 101) notes: 'If few artists outside the Western world have been avowed rebels or crusaders, many have exhibited an originality or independence that manifests the actual freedom of the human spirit, and that might weaken the despotism of custom and conventional belief.'
9. A cogent example of this expression of liberty through art is given by Beattie (2016).
10. The way in which 'art lies outside the surveillance of mission control' and can be used as a medium to express forms of liberty and unpredictable creativity is discussed, using personal experiences, by Beattie (2015a,b).
11. The function of art in promoting a culture of liberty is well demonstrated by the countless attempts to stymie artistic expression by authoritarian governments. This is most convincingly (almost entertainingly) demonstrated by Karl Hägelin, censor for the Habsburg government in the eighteenth century. He apparently banned the use of the words liberty, Enlightenment, and freedom on stage, and tyranny and despotism were not to be mentioned because of his fear that his own government would be associated with these practices (discussed in Fichtner, 2011).

CHAPTER 8
Educating the free citizen

1. Bacon (2021, p. 20). There is no record that he himself used the oft-quoted shorter version, 'knowledge is power' (*scientia potentia es*),

although Thomas Hobbes did so in the 1668 Latin version of *Leviathan*. This sentence is to be found in the modern English version as 'The sciences, are small power' (Hobbes, 2017, p. 71).

2. We wish to avoid an education system whose purpose is to shore up the power of an extraterrestrial authority, in other words to avoid the charge made by Price (1991, p. 138) on an education system that 'has been a contraction, not an enlargement, of the intellectual faculties, an injection of false principles hardening them in error, not a discipline enlightening and improving them'.
3. Milton Friedman states (2002, p. 86): 'A stable and democratic society is impossible without a minimum degree of literacy and knowledge on the part of most citizens and without widespread acceptance of some common set of values. Education can contribute to both.' David Friedman (1995, p. 94) makes a similar point about the need for education to avoid a moral vacuum that can tempt the response of using police action to make up for a failure to educate.
4. de Vigne (2016, pp. 148, 152) notes the potential for technological intervention in enhancing education.
5. A more revolutionary take on this idea was made by Bakunin (1970, p. 62), who called for science to be taught to the public on the basis of its inherently rational approaches to investigating the world. One does not have to be a fan of Bakunin to find the general argument plausible.
6. de Vigne (2015b). Also, in de Vigne (2015a), the suggestion is made that education should include emphasis on the social skills needed to cooperate successfully in such extremes (e.g. 'reflection, understanding and management of self, responsibility and accountability in the community'). These, de Vigne emphasises, would be learned alongside science subjects.
7. Quite rightly, these three words ('rule of law') alone can encourage entire books: for example, Sandoz (1993). However, these tomes illustrate the considerable complexity that civilisation has developed in even single concepts within liberty and how important it is to educate people in them.
8. Powers (2011).
9. The publication of Spinoza's philosophical work in the seventeenth century triggered an energised discussion on the limits of free

speech (Israel, 2011). Of course, some people sympathetic towards John Stuart Mill and similar thinkers might argue as a point of principle that all speech should be freely allowed, as it is merely speech, not action that directly infringes another person; but in Western societies today, we have a category of legally punishable 'hate speech', so clearly not everyone agrees with this, and the matter remains an important subject of discussion.

10. Bay (1958, p. 301).
11. Education in all its forms takes the edge off the temptation for tyrannous behaviour. As De Staël observed with characteristic clarity (2008, p. 752): 'Examine the adversaries of freedom in every country, you will find among them a few deserters from the camp of men of talent, but in general, you will see that the enemies of freedom are the enemies of knowledge and intelligence. They are proud of their deficiency in this respect; and one must agree that such a negative triumph can be easily achieved.'
12. The link between the type of state one creates and the virtue of its citizens, particularly their inherent traits of modesty, was also discussed by Nedham (1991, pp. 31–6).
13. Teenage liberty in space, and the problems of misappropriately directed youthful excess, has been examined by de Vigne (2016).
14. I have used the cage analogy previously: 'It would not be inaccurate to say simply that children born in space will be the first humans to be reared in cages' (Cockell, 2008, p. 267). I stick to it because I think that it captures an essential reality for youthful expression that a society needs to do its best to overcome through education.
15. McMahon (2016).

CHAPTER 9
Justice and criminality in the free society

1. Put simply by Mill (2001, p. 54): 'but security no human being can possibly do without; on it we depend for all our immunity from evil and for the whole value of all and every good, beyond the passing moment, since nothing but the gratification of the instant could

be of any worth to us if we could be deprived of anything the next instant by whoever was momentarily stronger than ourselves'.

2. Fukuyama (2012, p. 247). Hayek (2013) also gives considerable thought to the characteristics of legislation and the system of justice and the relationship between them (pp. 197–266), providing some insight into the complexity of these systems within democracies beyond the existence of the laws themselves. Note that his view on the complexity of the juridical aspects of society would still be relevant even if we conceived of a libertarian private legal system. That too still requires robust systems of law and their implementation, even if by private entities, to function in a way that commands the obedience of those that use it. See, for example, Benson (1990).
3. Of course, they were not the first. China's legalists also placed written laws at the apex of a well-ordered society about 2,000 years before, notably Han Feizi. See Black (2016, pp. 103–5 for a discussion).
4. Rosanvallon (2007, pp. 49–50). The context of this culture and some discussion of the consequences can be found in De Staël (2008).
5. These sets of conditions within the justice system have been identified as being necessary for societies to move out of authoritarian regimes as much as they are essential for retaining democratic societies (O'Donnell and Schmitter, 1990).
6. Russell (2004, p. 81) observes: 'The exercise of power, if it is to be something better than the infliction of wanton torture, must be hedged round by safeguards of law and custom, permitted only after due deliberation, and entrusted to men who are closely supervised in the interests of those who are subjected to them.'
7. The purpose of equality before the law is described by von Mises (2005, p. 10). It is not only that this allows predictability for the individual—knowing how they will be treated under the law—but also that there is likely to be a great deal of friction where different classes and groups are treated differently. And in an extraterrestrial society confined under lethal conditions, this simmering sense of injustice is dangerous.
8. Cockell (2008, p. 265).

9. Pettit regards this as a basic necessity in a republic, where 'domination' by arbitrary power is to be avoided: 'There should be procedures that ensure, for example, that the bureaucracy or the police or the courts cannot ignore certain challenges that are raised against them' (Pettit, 2010, p. 195).
10. Bay (1958, p. 151), writing about the future of the American police force, notes that: 'Whatever the future course of American law, it seems to me desirable that the police power should be used to enforce a maximum tolerance of free political speech and never to silence speakers just because intolerant individuals threaten to do so unlawfully.' The somewhat unfamiliar idea of police going around 'enforcing' more liberty can of course quickly become anything but desirable. However, the idea that the mentality of a police force should involve not only looking for instances of law-breaking, but actively doing things to enhance the expression of freedoms to the maximum extent possible has a lot to recommend it, even if only to ingrain the values of liberty into a police force.
11. von Humboldt (1993, pp. 86–93) proffered a vision of the limits of the police as focused only on the actions of individuals that infringe others without their consent or against their will. However, he did also allow for action when 'the probability of such consequences is to be feared'. In the extremities of the extraterrestrial environment, this could be interpreted as an excuse to take police action, given the potentially more lethal consequences of actions against, for example, the infrastructure. Even though von Humboldt did say that such actions should be circumscribed by what the potential consequences could be, if the potential consequences are depressurisation, then the police action could be draconian. Of course, there will be some level of self-regulation in a police force, since violent police action could bring disaster to a settlement as well as criminal activity itself. The settlers will likely need to demand a police force force whose practical capacity for the use of violence is carefully circumscribed to prevent law enforcement from causing catastrophe in confined spaces. Self-evidently, in the extraterrestrial environment, oversight of a police force is needed to correct

any overreach of power in these proactive circumstances of law enforcement.

12. I am assuming here the right to punish. Of course, this is in itself a matter that can be deliberated. For example, Schmidtz (1991) explores the basis of a state's presumed right to punish offenders.
13. Aside from their use in ancient Athens, many more recent isolated communities in extremes, such as the Inuit of the High Arctic, have used banishment and ostracism as a means of dealing with serious social transgressions (Fossett 2001). These means cannot work in the native extraterrestrial environment (at least if one subscribes to the most basic humane behaviour), but they could in theory work if an individual can be moved with dignity to a larger population that can take them in or provide better facilities for their management.
14. For example, Morris and Rothman (1997); Fairweather and McConville (2000); Brodie et al. (2002); Wener (2014). There is also the intriguing and important paper by Kutner (1968), which suggests the construction of extraterrestrial prisons to hold prisoners from Earth. He suggests building prisons in space to hold the whole world's prison population and remove them from Earth. This is a very different suggestion from mine, but it is, nevertheless, a rare paper that discusses space exploration and prisons. Kutner does not discuss the physical architecture or ethos of such a prison.
15. 'There can be no punishment more severe than the loss of one's freedom' was a sentiment expressed by an individual with prison experience: Boyle (1985, p. 34).
16. There are some people who think that we should attempt to do away with the institutions of prisons completely. In this work, I am not going to debate these very extensive arguments, but I am making the working assumption that an incarceration facility is required, even if it is to hold a person who is drunk and disorderly or suffering from short-term rage or any other condition that requires them to be restricted in their movement for some temporary period.
17. Discussed in Zimbardo (2008).

18. Stevens (2015) discusses the reduction in oxygen concentration in air mixtures as a possible form of coercion or even the formation of a 'class' system established by the richness of oxygen you can gain access to.
19. Cockell (2016c).
20. Prison population figures: https://www.gov.uk/government/statistics/prison-population-figures-2018. Retrieved on 16 February 2021.
21. Cockell et al. (2018).
22. The prisoners produced two books (Cockell, 2018, 2020). Perhaps extraterrestrial inmates might use their skills to design stations and parts of outposts as extensions of their own station or other extraterrestrial bases known to them.
23. There is an enormous literature on this topic, but see Vacca (2004) and Stevens and Ward (1997).

CHAPTER 10
A free cosmos

1. Thucydides (1972, p. 145).
2. Wollstonecraft (2001).

WORKS CITED

Note that the dates provided are for editions cited in the text, not necessarily the original date of publication. For works more than one hundred years old, for which the original is not the version used, the original date of publication is shown in square brackets to provide some chronological information for the accompanying notes.

ACEMOGLU, D., ROBINSON, J. A. (2020) *The Narrow Corridor: How Nations Struggle for Liberty*. London: Penguin Books.

ADAMS, J. (2014) [1765] *A Dissertation on the Canon and Feudal Law*. CreateSpace Independent Publishing Platform.

ADAMS, J. (2020) [The Defence of the Constitution: 1787] *The Works of John Adams. Vol. 4: Novanglus, Thoughts on Government, Defence of the Constitution*. Loschberg: Jazzybee Verlag.

ADLER, M. J. (1958) *The Idea of Freedom*. New York: Doubleday and Company.

AGARD, W. M. (2018) *What Democracy Meant to the Greeks*. London: Creative Media Partners, LLC.

ALCOTT, L. M. (2012) [1873] *Transcendental Wild Oats: A Chapter from an Unwritten Romance*. Print on demand: Forgotten Books.

ALEXANDER, D. (2017) *The Potter's Tale. A Colonsay Life*. Edinburgh: Birlinn.

ALFORD, C. F. (2005) *Rethinking Freedom: Why Freedom Has Lost Its Meaning and What Can Be Done to Save It*. New York: Palgrave Macmillan.

ALLEN, J. (2020) [1773] *An Oration on the Beauties of Liberty, or The Essential Rights of the Americans. Delivered at the Second Baptist-Church in Boston, Upon the Last Annual Thanksgiving, December 3, 1772*. Print on demand: Ecco Print Editions.

ALLEN, J. (2021) [1773] *The American alarm, or The Bostonian plea, for the rights, and liberties, of the people. Humbly addressed to the King and Council, and to the constitutional sons of liberty, in America. By the British Bostonian*. Print on demand: Hanse Books.

ALMÁR, I. (2002) What could COSPAR do to protect the planetary and space environment? *Advances in Space Research* 30, 1577–81.

Altman, L. J. (1996) *Life on an Israeli Kibbutz*. San Diego: Lucent Books.

Anshen, R. N. (1942) *Freedom: Its Meaning*. London: George Allen and Unwin.

Arendt, H. (1976) *The Origins of Totalitarianism*. San Diego: Harcourt.

Arendt, H. (2006) *Between Past and Future*. London: Penguin.

Arendt, H. (2016) *On Revolution*. London: Faber and Faber.

Aristotle (1988) [~350 bce] *The Politics*. London: Penguin.

Asma, S. T. (2018) *Why We Need Religion*. Oxford: Oxford University Press.

Bacon, F. (2008) [1626] *Three Early Modern Utopias*. Oxford: Oxford University Press.

Bacon, F. (2021) [1597] *Meditationes Sacrae*. Retrieved from https://en.wikisource.org/wiki/Meditationes_sacrae on 3 July 2021.

Bagehot, W. (2009) [1867] *The English Constitution*. Oxford: Oxford University Press.

Bailyn, B. (2017) *The Ideological Origins of the American Revolution*. Cambridge, MA: Harvard University Press.

Bakunin, M. (1970) [1882] *God and the State*. New York: Dover Publications.

Barkai, H. (1972) The kibbutz: An experiment in microsocialism. In *Israel, The Arabs and The Middle East* (eds: Howe, I. and Gershman, C), pp. 9–99. London: Bantam Books.

Baxter, S. (2015a) The cold equations: Extraterrestrial liberty in science fiction. In *The Meaning of Liberty Beyond Earth* (ed: Cockell, C. S.), pp. 13–31. Heidelberg: Springer.

Baxter, S. (2015b) The birth of a new republic: Depictions of the governance of a free Moon in science fiction. In *Human Governance Beyond Earth: Implications for Freedom* (ed: Cockell, C. S.), pp. 51–61. Heidelberg: Springer.

Baxter, S. (2016) Architects of the revolution: Design philosophies to maximise extraterrestrial liberty. In *Dissent, Revolution and Liberty Beyond Earth* (ed: Cockell, C. S.), pp. 111–23. Heidelberg: Springer.

Bay, C. (1958) *The Structure of Freedom*. Stanford: Stanford University Press.

Beattie, A. (2015a) Extraterrestrial liberty and creative practice: A first-hand experience of an imagined future. In *The Meaning of Liberty Beyond Earth* (ed: Cockell, C. S.), pp. 103–16. Heidelberg: Springer.

BEATTIE, A. (2015b) Art making, good governance and sustainable extraterrestrial communities. In *Human Governance Beyond Earth: Implications for Freedom* (ed: Cockell, C. S.), pp. 213–24. Heidelberg: Springer.

BEATTIE, A. (2016) Art and change in an extraterrestrial society. In *Dissent, Revolution and Liberty Beyond Earth* (ed: Cockell, C. S.), pp. 133–42. Heidelberg: Springer.

BEDAU, H. (1961) On civil disobedience. *Journal of Philosophy* 58, 661.

BEDAU, H. (1970) Civil disobedience and personal responsibility for injustice. *The Monist* 54, 517–35.

BENEKE, C. (2006) *Beyond Toleration: The Religious Origins of American Pluralism.* Oxford: Oxford University Press.

BENSON, B. L. (1990) *The Enterprise of Law.* San Francisco: Pacific Research Institute for Public Policy.

BERLIN, I. (1979) *Against the Current: Essays in the History of Ideas.* London: Pimlico.

BERLIN, I. (1998) *The Proper Study of Mankind.* London: Pimlico.

BERLIN, I. (1999) *The Roots of Romanticism.* London: Pimlico.

BERLIN, I. (2003) *Freedom and Its Betrayal: Six Enemies of Human Liberty.* Princeton, NJ: Princeton University Press.

BERLIN, I. (2006) *The Crooked Timber of Humanity.* London: Pimlico.

BERLIN, I. (2008) *Russian Thinkers.* London: Penguin.

BHATT, M. C. (2015) Constituting outer space: The governance of planetary settlements and artificial habitats. In *Human Governance Beyond Earth: Implications for Freedom* (ed: Cockell, C. S.), pp. 149–71. Heidelberg: Springer.

BHATT, M. C. (2016) Space for dissent: Disobedience on artificial habitats and planetary settlements. In *Dissent, Revolution and Liberty Beyond Earth* (ed: Cockell, C. S.), pp. 71–92. Heidelberg: Springer.

BLACK, A. (2016) *A World History of Political Thought: Its Significance and Consequences.* Oxford: Oxford University Press.

BOLINGBROKE, VISCOUNT (1997) [eighteenth-century letters and other writings] *Political Writings.* Cambridge: Cambridge University Press.

BOOKCHIN, M. (2005) *The Ecology of Freedom.* Edinburgh, UK: AK Press.

BOROWIAK, C. T. (2011) *Accountability and Democracy.* Oxford: Oxford University Press.

BOSANQUET, B. (1920) *The Philosophical Theory of the State.* London: Macmillan.

BOTTON DE, A. (2014) *The Architecture of Happiness.* London: Penguin Books.

BOYLE, J. (1985) *The Pain of Confinement: Prison Diaries.* London: Pan Books.

BRENKERT, G. G. (1991) *Political Freedom.* London: Routledge.

BREYER, S. (2008) *Active Liberty: Interpreting a Democratic Constitution.* Oxford: Oxford University Press.

BRODIE, A., CROOM, J., DAVIES, J. O. (2002) *English Prisons: An Architectural History.* London: English Heritage.

BROWN, R. D. (1995) The idea of an informed citizenry in the early republic. In *Devising Liberty: Preserving and Creating Freedom in the New American Republic* (ed: Konig, T.), pp. 141–77. Stanford: Stanford University Press.

BRUHNS, S., HAQQ-MISRA, J. (2016) A pragmatic approach to sovereignty on Mars. *Space Policy* 38, 57–63.

BRYCE, J. (1908) *The American Commonwealth.* Vol. 2. London: Macmillan.

BUCHANAN, J. M. (1975) *The Limits of Liberty: Between Anarchy and Leviathan.* Chicago: Chicago University Press.

BURKE, E. (1986) [1790] *Reflections on the Revolution in France.* London: Penguin Books.

BURY, J. B. (1952) *A History of Freedom of Thought.* New York: International Debate Education Association.

BURY, J. B. (2006) *The Idea of Progress. An Inquiry into Its Origin and Growth.* Fairford: Echo Library.

CAIN, J. R. (2015a) The Marxist theory of capital as applied to the economic development and governance of space colonies. In *Human Governance Beyond Earth: Implications for Freedom* (ed: Cockell, C. S.), pp. 81–101. Heidelberg: Springer.

CAIN, J. R. (2015b) Astronaut health: Exploration and the limitations on freedom. In *The Meaning of Liberty Beyond Earth* (ed: Cockell, C. S.), pp. 139–63. Heidelberg: Springer.

CAIN, J. R. (2016) Space terrorism: A new environment; new causes. In *Dissent, Revolution and Liberty Beyond Earth* (ed: Cockell, C. S.), pp. 93–109. Heidelberg: Springer.

CARTER, I. (2004) *A Measure of Freedom.* Oxford: Oxford University Press.

CATLING, D. C., GLEIN, C. R., ZAHNLE, K. J., MCKAY, C. P. (2005) Why O_2 is required by complex life on habitable planets and the concept of planetary 'oxygenation time'. *Astrobiology* 5, 415–38.

CATO'S LETTERS (1995) [1720] *Essays on Liberty, Civil and Religious, and Other Important Subjects*. Indianapolis: Liberty Fund.

CHESTERTON, G. K. (1929) *The Thing*. London: Sheed and Ward.

CHRISTAKIS, N. A. (2019) *Blueprint: The Evolutionary Origins of a Good Society*. New York: Little, Brown Spark.

CICERO (2003) [44 BCE] *On Duties*. Cambridge: Cambridge University Press.

CICERO (2008) [~50 BCE] *The Republic and The Laws*. Oxford: Oxford University Press.

CLARK, J. C. D. (1994) *The Language of Liberty 1660–1832*. Cambridge: Cambridge University Press.

COCKELL, C. S. (2002) The Trans-Mars Expedition: A long-distance, long-duration, scientific EVA. *Journal of the British Interplanetary Society* 55, 291–306.

COCKELL, C. S. (2008) An essay on extraterrestrial liberty. *Journal of the British Interplanetary Society* 61, 255–75.

COCKELL, C. S. (2009) Liberty and the limits to the extraterrestrial state. *Journal of the British Interplanetary Society* 61, 139–57.

COCKELL, C. S. (2010) Essay on the causes and consequences of extraterrestrial tyranny. *Journal of the British Interplanetary Society* 63, 15–37.

COCKELL, C. S. (2011) Liberty across light years: The freedom of space settlers compared to that of the ancients and the moderns. *Journal of the British Interplanetary Society* 64, 287–88.

COCKELL, C. S. (editor) (2015a) *The Meaning of Liberty Beyond Earth*. Heidelberg: Springer.

COCKELL, C. S. (2015b) Freedom in a box: Paradoxes in the structure of extraterrestrial liberty. In *The Meaning of Liberty Beyond Earth* (ed: Cockell, C. S.), pp. 47–68. Heidelberg: Springer.

COCKELL, C. S. (2015c) Extraterrestrial liberty: Can it be planned? In *Human Governance Beyond Earth: Implications for Freedom* (ed: Cockell, C. S.), pp. 23–41. Heidelberg: Springer.

COCKELL, C. S. (2015d) The purpose of art and artists beyond Earth. *Journal of the British Interplanetary Society* 68, 376–82.

Cockell, C. S. (editor) (2016a) *Dissent, Revolution and Liberty Beyond Earth*. Heidelberg: Springer.

Cockell, C. S. (2016b) Disobedience in outer space. In *Dissent, Revolution and Liberty Beyond Earth* (ed: Cockell, C. S.), pp. 21–40. Heidelberg: Springer.

Cockell, C. S. (2016c) Exoconfac—The extraterrestrial containment facility: An essay on the design philosophy for a prison to contain criminals in settlements beyond the Earth. *Journal of the British Interplanetary Society* 69, 27–30.

Cockell, C. S. (editor) (2018) *Life Beyond: From Prison to Mars*. London: British Interplanetary Society.

Cockell, C. S. (2019) Freedom engineering: Using engineering to mitigate tyranny in space. *Space Policy* 49, article no. 101328.

Cockell, C. S. (editor) (2020) *Life Beyond 2: From Prison to the Moon*. London: British Interplanetary Society.

Cockell, C. S., Fosado, Y. A. G., Hitchen, J., Landenmark, H., Perera, L., Vissers, T. (2018) Life Beyond: A program to use astrobiology to teach science and advance space exploration through prisons. *Journal of Correctional Education* 69, 30–43.

Cockell, C. S., Horneck, G. (2004) A Planetary Park system for Mars. *Space Policy* 20, 291–5.

Cockell, C. S., Horneck, G. (2006) Planetary Parks: Formulating a wilderness policy for other planetary bodies. *Space Policy* 22, 256–61.

Cohn, S. K. (2008) *Lust for Liberty: The Politics of Social Revolt in Medieval Europe, 1200–1425*. Cambridge, MA: Harvard University Press.

Coke, E. (2002) [collection from sixteenth and early seventeenth centuries] *The Selected Writings of Sir Edward Coke*. Indiana: Liberty Fund.

Colbourn, T. (1998) *The Lamp of Experience: Whig History and the Intellectual Origins of the American Revolution*. Indiana: Liberty Fund.

Condorcet de, A.-N. (1955) [1795] *Sketch for a Historical Picture of the Progress of the Human Mind*. Westport: Hyperion Press.

Cowles, H. M. (2020) *The Scientific Method: An Evolution of Thinking from Darwin to Dewey*. Cambridge, MA: Harvard University Press.

Cowley, R. (2019) Yes, we Earthlings should colonize Mars if Martian rights can be upheld. *Theology and Science* 17, 332–40.

CRAWFORD, I. A. (2015) Interplanetary federalism: Maximising the chances of extraterrestrial peace, diversity and liberty. In *The Meaning of Liberty Beyond Earth* (ed: Cockell, C. S.), pp. 199–218. Heidelberg: Springer.

CRAWFORD, I. A., BAXTER, S. (2015) The lethality of interplanetary warfare: A fundamental constraint on extraterrestrial liberty. In *The Meaning of Liberty Beyond Earth* (ed: Cockell, C. S.), pp. 187–98. Heidelberg: Springer.

CROUCH, C. (1982) *The Logic of Collective Action*. London: Fontana.

DAHL, R. A. (2020) *On Democracy*. New Haven, CT: Yale University Press.

DAVIES, J. K. (1993) *Democracy and Classical Greece*. London: Fontana.

DELMAS, C. (2018) *A Duty to Resist: When Disobedience Should Be Uncivil*. Oxford: Oxford University Press.

DE LOLME, J. L. (2007) [1771] *The Constitution of England*. Indianapolis: Liberty Fund.

DE STAËL, G. (2008) [1818] *Considerations on the Principal Events of the French Revolution*. Indianapolis, IN: Liberty Fund.

DE TOCQUEVILLE, A. (1998) [1835] *Democracy in America*. London: Bibliophile Books.

DE TOCQUEVILLE, A. (2008) [1856] *The Ancien Régime and the Revolution*. London: Penguin.

DELANO, S. F. (2004) *Brook Farm: The Dark Side of Utopia*. Cambridge, MA: Harvard University Press.

DELMAS, C. (2018) *A Duty to Resist: When Disobedience Should be Uncivil*. Oxford: Oxford University Press.

DEUDNEY, D. (2020) *Dark Skies: Space Expansionism, Planetary Geopolitics, and the Ends of Humanity*. Oxford: Oxford University Press.

DEWEY, R. E., GOULD, J. A, (1970) *Freedom: Its History, Nature, and Varieties*. London: Macmillan Company.

DICKINSON, J. (2020) [1769] *Letters from a Farmer in Pennsylvania, to the Inhabitants of the British Colonies*. Print on demand: Ecco Print Editions.

DOLMAN, E. C. (2002) *Astropolitik: Classical Geopolitics in the Space Age*. London: Frank Cass.

DUNN, J. (1992) *Democracy: The Unfinished Journey*. Oxford: Oxford University Press.

EDWARDS, S. (1971) *The Paris Commune 1871*. Newton Abbot: Eyre and Spottiswoode.

ELVIS, M. (2013) How many ore-bearing asteroids? *Planetary and Space Science* 91, 20–6.

ELVIS, M., KROLIKOWSKI, A., MILLIGAN, T. (2021) Concentrated lunar resources: Imminent implications for governance and justice. *Philosophical Transactions of the Royal Society A* 379, 20190563.

ESBECK, C. H., DEN HARTOG, J. J. (2019) *Disestablishment and Religious Dissent: Church-state Relations in the New American States 1776–1833*. Columbia, MO: University of Missouri Press.

FAIRWEATHER, L. F., MCCONVILLE, S. (2000) *Prison Architecture*. Abingdon: Routledge.

FERGUSON, A. (2007) [1767] *An Essay on the History of Civil Society*. Fairford: Echo Library.

FERL, F., WHEELER, R., LEVINE, H. G., PAUL, A. L. (2002) Plants in space. *Current Opinions in Plant Biology* 5, 258–63.

FERRIS, T. (2010) *The Science of Liberty*. New York: HarperCollins.

FICHTNER, P. S. (2011) Print vs. speech: Censoring the stage in eighteenth century Vienna. In *Freedom of Speech: The History of an Idea*. (ed: Powers, E.), pp. 81–101. Lewisburg: Bucknell University Press.

FILMER, R. (1991) [1680] *'Patriarcha' and Other Writings*. Cambridge: Cambridge University Press.

FINLAY, C. (2000) *Art of the Far North: Inuit Sculpture, Drawing and Printmaking*. Minneapolis, MN: Lerner Publications

FINLAY, M. I. (1985) *Democracy: Ancient and Modern*. New Brunswick: Rutgers University Press.

FOGG, M. J. (1995) *Terraforming: Engineering Planetary Environments*. Warrendale, PA: SAE International.

FORGAN, D. H. (2011) Spatio-temporal constraints on the zoo hypothesis, and the breakdown of total hegemony. *International Journal of Astrobiology* 10, 341–7.

FOSSETT, R. (2001) *In Order to Live Untroubled: Inuit of the Central Arctic, 1550–1940*. Manitoba: University of Manitoba Press.

FRANKL, V. E. (2004) *Man's Search for Meaning*. London: Penguin.

FRIEDMAN, D. (1995) *The Machinery of Freedom*. Chicago: Open Court.

FRIEDMAN, M. (2002) *Capitalism and Freedom.* Chicago: University of Chicago Press.

FRIEDMAN, M. (2017) *Milton Friedman on Freedom.* Stanford: Hoover Institution Press.

FROMM, E. (2006) *The Fear of Freedom.* London: Routledge.

FUKUYAMA, F. (2012) *The Origins of Political Order.* London: Profile Books.

FULLER, L. L. (1965) Irrigation and tyranny. *Stanford Law Review* 17, 1021–42.

GANDHI, M. (1961) *Nonviolent Resistance.* New York: Schocken Books.

GOWER, B. (1996) *Scientific Method: A Historical and Philosophical Introduction.* London: Routledge.

GRAY, J. (1991) *Liberalisms: Essays in Political Philosophy.* London: Routledge.

GRAYLING, A. C. (2009) *Liberty in the Age of Terror.* London: Bloomsbury.

GRAZIOSI, D., LEE, R. (2003) *I-suit Advanced Spacesuit Design Improvements and Performance Testing,* SAE technical paper, 2003-01-2443.

GREENALL-SHARP, R., KOBZA, D., HOUSTON, C., ALLABBAD, M., STAGGS, J., SCHWARTZ, J. S. J. (2021) A space settler's Bill of Rights. In *Astrobiology: Science, Ethics, and Public Policy* (eds: Chon Torres, O. A., Peters, T.), pp. 377–88. Hoboken, NJ: John Wiley and Sons, Inc.

GRONSTAL, A. L., PEARSON, V., KAPPLER, A., ANAND, M., POITRASSON, F., KEE, T. P., COCKELL, C. S. (2009) Laboratory experiments on the weathering of iron meteorites and carbonaceous chondrites by iron-oxidising bacteria. *Meteoritics and Planetary Science* 44, 233–48.

GUERRIERI, M. L., LEWIS, J. S., MATTHEWS, M. S. (1993) *Resources of Near-Earth Space.* Tucson: University of Arizona Press.

HAMILTON, A. (2010) [1788] The Federalist No. 76. In *The Federalist with Letters of 'Brutus',* pp. 371–2. Cambridge: Cambridge University Press.

HANSEN, M. H. (2006) *Polis: An Introduction to the Ancient Greek City-State.* Oxford: Oxford University Press.

HAQQ-MISRA, J. (2015) Towards a sustainable land use policy for Mars. In *Human Governance Beyond Earth: Implications for Freedom* (ed: Cockell, C. S.), pp. 43–9. Heidelberg: Springer.

HAQQ-MISRA, J. (2016) The transformative value of liberating Mars. *New Space* 4, 64–7.

HAQQ-MISRA, J. (2019) Can deep altruism sustain space settlement? In *The Human Factor in a Mission to Mars* (ed: Szocik, K.), pp. 145–55. Heidelberg: Springer.

HARASZTI, Z. (1952) *John Adams and The Prophets of Progress*. Cambridge, MA: Harvard University Press.

HARRIS, P. H. (2009) *Space Enterprise: Living and Working Offworld in the 21st Century*. Heidelberg: Springer/Praxis.

HÄUPLIK-MEUSBURGER, S., BISHOP S. (2021) *Space Habitats and Habitability*. Heidelberg: Springer.

HAYEK, F. A. (1980) *Individualism and Economic Order*. Chicago: University of Chicago Press.

HAYEK, F. A. (2006) *The Constitution of Liberty*. London: Routledge.

HAYEK, F. A. (2007) *The Road to Serfdom*. London: Routledge.

HAYEK, F. A. (2013) *Law, Legislation and Liberty*. London: Routledge.

HEADLAM, J. W. (1890) *Election by Lot at Athens*. London: Forgotten Books.

HECHT, M. H., KOUNAVES, S. P., QUINN, R. C., WEST, S. J., YOUNG, S. M. M., MING, D. W., CATLING, D. C., CLARK, B. C., BOYNTON, W. V., HOFFMAN, J., DEFLORES, L. P., GOSPODINOVA, K., KAPIT, J., SMITH, P. H. (2009) Detection of perchlorate and the soluble chemistry of Martian soil at the Phoenix lander site. *Science* 325, 64–7.

HEINLEIN, R. A. (2005) *The Moon is a Harsh Mistress*. London: Hodder Paperbacks.

HERBERT, A. (1978) *The Right and Wrong of Compulsion by the State: And Other Essays by Auberon Herbert*. Indianapolis: Liberty Press.

HESSEL, I. (2006) *Arctic Spirit: Inuit Art from the Albrecht Collection at the Heard Museum*. Madeira Park, BC: D&M Publishers Inc.

HIGGS, R. (2012) *Crisis and Leviathan*. Oakland: The Independent Institutes.

HOBBES, T. (2017) [1651] *Leviathan*. London: Penguin Books.

HOBHOUSE, L.T. (2009) [1911] *Liberalism*. Fairford: Echo Library.

HOGAN, C. (2021) Extraterrestrial netizenship: Citizenship in outer space and cyberspace. In *Outer Space and Cyberspace* (ed: Froehlich, A.), pp. 15–28. Heidelberg: Springer.

HOPKINS, M. M. (1979) *The Economics of Strikes and Revolts During Early Space Colonization: A Preliminary Analysis.* Santa Monica, CA: RAND Corporation.

HUEMER, M. (2013) *The Problem of Political Authority. An Examination of the Right to Coerce and the Duty to Obey.* New York: Palgrave Macmillan.

HUMBOLDT, W. VON (1993) [1852] *The Limits of State Action.* Indianapolis, IN: Liberty Fund.

HUME, D. (1777) *Idea of a Perfect Commonwealth.* Retrieved from https://oll.libertyfund.org/page/oll-reader-70 on 14 November 2020.

INGLIS, C. (1776) *The True Interest of America Impartially Stated, in Certain Strictures on a Pamphlet Intitled Common Sense: By an American.* Retrieved from https://quod.lib.umich.edu/e/evans/N11718.0001.001?rgn=main;view=fulltext on 12 February 2021.

ISRAEL, J. I. (2011) *Libertas Philosophandi* in the eighteenth century: Radical enlightenment versus moderate enlightenment (1750–1766). In *Freedom of Speech: The History of an Idea.* (ed: Powers, E.), pp. 1–17. Lewisburg, PA: Bucknell University Press.

JAFFA, H. V. (2004) *A New Birth of Freedom: Abraham Lincoln and the Coming of the Civil War.* London: Rowman and Littlefield.

JEFFERSON, T. (1787) Letter from Thomas Jefferson to James Madison, January 30, 1787. Retrieved from https://www.loc.gov/resource/mtj1.006_1110_1119/ on 12 February 2021.

JOHNSON-FREESE, J. (2007) *Space as a Strategic Asset.* New York: Columbia University Press.

JONES, A. H. M. (1977) *Athenian Democracy.* Baltimore, MD: Johns Hopkins University Press.

JORDAN, N., SALEH, J., NEWMAN, D. (2006) The extravehicular mobility unit: A review of environment, requirements, and design changes in the US spacesuit. *Acta Astronautica* 59, 1135e1145.

JOUVENEL DE, B. (1993) *On Power: The Natural History of Its Growth.* Indianapolis, IN: Liberty Fund.

KANT, I. (2012) [1785] *Groundwork of the Metaphysics of Morals.* Cambridge: Cambridge University Press.

KASSON, J. F. (1999) *Civilizing the Machine: Technology and Republican Values in America, 1776–1900.* New York: Hill and Wang.

KEANE, J. (2002) Cosmocracy. A global system of governance or anarchy? *New Economy* 9, 65–70.

KLOSKO, G. (2010) *Political Obligations.* Oxford: Oxford University Press.

KUHN, T. S. (2012) *The Structure of Scientific Revolutions.* Chicago, IL: Chicago University Press.

KUTNER, L. (1968) A world outer space prison. *Denver Law Review* 45, 702–18.

LAUDANI, R. (2013) *Disobedience in Western Political Thought.* Cambridge: Cambridge University Press.

LE BON, G. (2005) [1896] *The Crowd: A Study of the Popular Mind.* Minneapolis: Filiquarian Publishing LLC.

LENIN, V. (1932) [1917] *The State and Revolution.* New York: International Publishers.

LEVY, J. T. (2015) *Rationalism, Pluralism, and Freedom.* Oxford: Oxford University Press.

LEWIS, J. S. (1997) *Mining the Sky.* Basic Books: New York.

LIM, J., CHRISTENSEN, R. (2020) *Thriving Not Just Surviving: The Right to a Breathable Atmosphere.* Jus Ad Astra project (July 2020).

LIN, P. (2006) Viewpoint: Look before taking another leap for mankind: Ethical and social considerations in rebuilding society in space. *Astropolitics* 4, 281–94.

LIVY (1982) [~30 BCE to 15 CE] *Rome and Italy.* London: Penguin Books.

LOCKE, J. (1988) [1689] *Two Treatises of Government.* Cambridge: Cambridge University Press.

LOVETT, F. (2010) *A General Theory of Domination and Justice.* Oxford: Oxford University Press.

LUCAS-RHIMBASSEN, M. (2021) On the province of all Marskind. In *Assessing a Mars Agreement Including Human Settlements* (ed: Froehlich, A.), pp. 27–36. Heidelberg: Springer.

LUKES, S. (2005) *Power: A Radical View.* London: Palgrave Macmillan.

MACAULAY, C. (2018) [1770] *Observations on a Pamphlet, Entitled, Thoughts on the Cause of the Present Discontents.* Print on demand: Hanse Books.

MACHIAVELLI, N. (2008) [1531] *Discourses on Livy.* Oxford: Oxford University Press.

MACKLEM, T. (2006) *Independence of Mind.* Oxford: Oxford University Press.

MacLean, C. (2006) *Island on the Edge of the World. The Story of St Kilda.* Canongate: Edinburgh.

Madison, J. (1788) Virginia Ratifying Constitution. Papers 11: 163 (June 10 1788). Retrieved from https://press-pubs.uchicago.edu/founders/documents/v1ch13s36.html on 11 January 2021.

Madison, J. (2010) [1788] The Federalist No. 51. In *The Federalist with Letters of 'Brutus'*, pp. 251–5. Cambridge: Cambridge University Press.

Malinowski, B. (1960) *Freedom and Civilization.* Bloomington, IN: Indiana University Press.

Mandeville, B. (1988) [1714] *The Fable of the Bees, or Private Vices, Publick Benefits. Volumes I and II.* Indianapolis, IN: Liberty Fund.

Manin, B. (1997) *The Principles of Representative Government.* Cambridge: Cambridge University Press.

Marx, K. (1966) [1891] *Critique of the Gotha Program.* New York: International Publishers Co., Inc.

Mason, C. R. (2021) *The Next 500 Years: Engineering Life to Reach New Worlds.* Cambridge, MA: MIT Press.

Matthews, J. J., McMahon, S. (2018) Exogeoconservation: Protecting geological heritage on celestial bodies. *Acta Astronautica* 149, 55–60.

Mautner, M. N. (1997) Biological potential of extraterrestrial materials—1. Nutrients in carbonaceous meteorites, and effects on biological growth. *Planetary and Space Science* 45, 653–61.

Mayhew, J. (2020) [1751] *A Discourse, Concerning Unlimited Submission and Non-resistance to the Higher Powers.* Washington, DC: Library of Congress.

McClellan, J. (2000) *Liberty, Order, and Justice.* Indianapolis, IN: Liberty Fund.

McIlwain, C. H. (2007) *Constitutionalism: Ancient and Modern.* Indianapolis, IN: Liberty Fund.

McKay, C. P., Toon, O. B., Kasting, J. F. (1991) Making Mars habitable. *Nature* 352, 489–96.

McKay, C. P., Marinova, M. (2001) The physics, biology, and environmental ethics of making Mars habitable. *Astrobiology* 1, 89–109.

McKay, C. P., Toon, O. B., Kasting, J. F. (1991) Making Mars habitable. *Nature* 352, 489–96.

McKnight, J. C. (2015) Space polities: Self-governance lessons from virtual worlds. In *Human Governance Beyond Earth: Implications for Freedom* (ed: Cockell, C. S.), pp. 103–19. Heidelberg: Springer.

McMahon, S. (2016) The aesthetic objection to terraforming. In *The Ethics of Space Exploration* (eds: Schwartz, J., Milligan, T.), pp. 209–18. Heidelberg: Springer.

Menger, C. (2009) [1892] *The Origins of Money*. Auburn, AL: Ludwig von Mises Institute.

Michels, R. (1962) [1911] *Political Parties. A Sociological Study of the Oligarchical Tendencies of Modern Democracy*. New York: The Free Press.

Mill, J. S. (1835) The state of the nation. *The London Review* 25, 1–24 (Mill signed off the paper 'P.Q.', as he did with some other papers).

Mill, J. S. (1998) [1859] *On Liberty and Other Essays* (ed: Gray, J.). Oxford: Oxford University Press.

Mill, J. S. (2001) [1863] *Utilitarianism*. Indianapolis, IN: Hackett Publishing Company.

Millar, J. (2006) [1779] *The Origin of the Distinction of Ranks*. Indianapolis, IN: Liberty Fund.

Miller, D. (1991) *Liberty*. Oxford: Oxford University Press.

Milligan, T. (2013) *Civil Disobedience: Protest, Justification, and the Law*. New York: Bloomsbury.

Milligan, T. (2015a) *Nobody Owns the Moon: The Ethics of Space Exploration*. Jefferson, NC: McFarland and Company, Inc.

Milligan, T. (2015b) Fear of freedom: The legacy of Arendt and Ballad's space skepticism. In *The Meaning of Liberty Beyond Earth* (ed: Cockell, C. S.), pp. 33–45. Heidelberg: Springer.

Milligan, T. (2015c) Rawlsian deliberation about space settlement. In *Human Governance Beyond Earth: Implications for Freedom* (ed: Cockell, C. S.), pp. 9–22. Heidelberg: Springer.

Milligan, T. (2016) Constrained dissent and the rights of future generations. In *Dissent, Revolution and Liberty Beyond Earth* (ed: Cockell, C.S.), pp. 7–20. Heidelberg: Springer.

Milton, J. (2014) [1644] *Areopagitica and Other Writings*. London: Penguin.

Mises, L. von (1944) *Omnipotent Government*. New Haven, CT: Yale University Press.

MISES, L. VON (1990) *Economic Freedom and Interventionism*. Indianapolis, IN: Liberty Fund.

MISES, L. VON (2005) *Liberalism*. Indianapolis, IN: Liberty Fund.

MISES, L. VON (2006) *The Anti-capitalistic Mentality*. Indianapolis, IN: Liberty Fund.

MISES, L. VON (2007) *Bureaucracy*. Indianapolis, IN: Liberty Fund.

MOLLER, D. (2019) *Governing Least: A New England Libertarianism*. Oxford: Oxford University Press.

MONTESQUIEU (2009) [1748] *The Spirit of the Laws*. Cambridge: Cambridge University Press.

MORLEY, F. (2010) *The Power in the People*. New Brunswick: Transaction Publishers.

MORRIS, C. (2002) *An Essay on the Modern State*. Cambridge: Cambridge University Press.

MORRIS, N., ROTHMAN, D. J. (1997) *The Oxford History of the Prison: The Practice of Punishment in Western Society*. Oxford: Oxford University Press.

MORTIMER, S. (2021) *Reformation, Resistance, and Reason of State (1517–1625)*. Oxford: Oxford University Press.

MULLER, H. J. (1960) *Issues of Freedom: Paradoxes and Promises*. New York: Harper and Brothers.

MULLER, H. J. (1961) *Freedom in the Ancient World*. New York: Harper and Brothers.

MULLER, H. J. (1963) *Freedom in the Western World*. New York: Harper and Brothers.

MULLER, H. J. (1966) *Freedom in the Modern World*. New York: Harper and Brothers.

MURPHY, A. R. (2018) *Liberty, Conscience and Toleration: The Political Thought of William Penn*. Oxford: Oxford University Press.

MUSK, E. (2017) Making humans a multi-planetary species. *New Space* 5, 46–61.

NASA TechPort—Mars OXygen ISRU Experiment Project (2018) NASA TechPort, National Aeronautics and Space Administration. Retrieved from https://mars.nasa.gov/mars2020/spacecraft/instruments/moxie/ on 12 November 2018.

NECKER, J. (2020) [1792] *On Executive Power in Great States.* Carmel, IN: Liberty Fund.

NEDHAM, M. (1991) [1656] *The Excellencie of a Free-State.* Indianapolis, IN: Liberty Fund.

NEHRU, J. (2010) *The Discovery of India.* London: Penguin Books.

NELSON, P. L., BLOCK, W. E. (2018) *Space Capitalism: How Humans Will Colonise Planets, Moons, and Asteroids.* London: Palgrave Macmillan.

NEUFELD, M. J. (1999) *The Rocket and the Reich.* Cambridge, MA: Harvard University Press.

NEWELL, W. R. (2013) *Tyranny: A New Interpretation.* Cambridge: Cambridge University Press.

NICHOLSON, W. L. SCHUERGER, A. C., RACE, M. S. (2009) Migrating microbes and planetary protection. *Trends in Microbiology* 17, 389–92.

NIEBUHR, R. (2008) *The Irony of American History.* Chicago, IL: Chicago University Press.

NISBET, R. A. (1962) *Community and Power.* New York: Galaxy Books.

NOZICK, R. (1974) *Anarchy, State, and Utopia.* Oxford: Blackwell Publishing.

NUSSBAUM, M. (2005) *Libertarianism: For and Against.* London: Rowman and Littlefield Publishers.

O'DONNELL, G., SCHMITTER, P. C. (1990) *Transitions from Authoritarian Rule: Tentative Conclusions about Uncertain Democracies.* Baltimore, MD: Johns Hopkins University Press.

O'NEILL, G. K. (1976) *The High Frontier: Human Colonies in Space.* New York: William Morrow and Company.

OPPENHEIM, F. E. (1961) *Dimensions of Freedom.* New York: St Martin's Press.

OSTROM, E. (2015) *Governing the Commons: The Evolution of Institutions for Collective Action.* Cambridge: Cambridge University Press.

OSTROM, V. (2008) *The Political Theory of a Compound Republic: Designing the American Experiment.* Lanham: Lexington Books.

OTIS, J. (2020) [1764] *The Rights of the British Colonies Asserted and Proved.* Print on demand: Ecco Print Editions.

OWEN, R. (1991) *A New View of Society and Other Writings.* London: Penguin.

PAINE, T. (1999) [1791] *The Rights of Man.* New York: Dover Publications.

PAINE, T. (2004) [1776] *Common Sense*. London: Penguin Books.

PALEY, W. (1821) [1785] *The Principles of Moral and Political Philosophy*. London: Baldwyn and Co.

PALMER, T. G. (2009) *Realizing Freedom. Libertarian Theory, History, and Practice*. Washington, DC: Cato Institute.

PASSMORE, J. (2000) *The Perfectability of Man*. Indianapolis, IN: Liberty Fund.

PATTERSON, O. (1991) *Freedom: Freedom in the Making of Western Culture*. New York: Basic Books.

PERSSON, E. (2015) Citizens of Mars Ltd. In *Human Governance Beyond Earth: Implications for Freedom* (ed: Cockell, C. S.), pp. 121–37. Heidelberg: Springer.

PETTIT, P. (2010) *Republicanism: A Theory of Freedom and Government*. Oxford: Oxford University Press.

PLATO (2008) [~375 BCE] *The Republic*. Oxford: Oxford University Press.

PLUTARCH (2005) [after 46 CE] *On Sparta*. London: Penguin.

POLYBIUS (2010) [written sometime between ~200 and 120 BCE] *The Histories*. Oxford: Oxford University Press.

POPPER, K. (1995) *The Open Society and Its Enemies*. London: Routledge.

POTTER, H. (2015) *Law, Liberty and the Constitution*. Woodbridge: The Boydell Press.

POWERS, E. (editor) (2011) *Freedom of Speech: The History of an Idea*. Lewisburg, MO: Bucknell University Press.

PRICE, R. (1991) [various eighteenth-century essays] *Political Writings*. Cambridge: Cambridge University Press.

PRIESTLEY, J. (1993) [various eighteenth-century essays] *Political Writings*. Cambridge: Cambridge University Press.

RADFORD, R. A. (1945) The economic organisation of a P.O.W. camp. *Economia* 12, 189–201.

RAPP, D. (2019) *Use of Extraterrestrial Resources for Human Space Missions to Moon or Mars*. Heidelberg: Springer.

RAWLS, J. (1971) *A Theory of Justice*. Cambridge, MA: Belknap Press of Harvard University Press.

RAWLS, J. (2008) *Lectures on the History of Political Philosophy*. Cambridge, MA: Belknap Press of Harvard University Press.

RAZ, J. (1988) *The Morality of Freedom*. Oxford: Clarendon Press.

REECE, E. (2016) *Utopia Drive: A Road Trip Through America's Most Radical Idea*. New York: Farrar, Straus and Giroux.

ROBERTS, J. L. (1994) *Athens on Trial*. Princeton, NJ: Princeton University Press.

ROBINSON, G. S., WHITE, H. M. (1986) *Envoys of Mankind*. Washington, DC: Smithsonian University Press.

ROGERS, T. F. (2004) Safeguarding Tranquility Base: Why the Earth's moon base should become a world heritage site. *Space Policy* 20, 5–6.

RÖPKE, W. (2008) *Economics of a Free Society*. Auburn, AL: Ludwig von Mises Institute.

ROSANVALLON, P. (2007) *The Demands of Liberty: Civil Society in France since the Revolution*. Cambridge, MA: Harvard University Press.

ROSHWALD, M. (2000) *Liberty: Its Meaning and Scope*. Westport, CT: Greenwood Press.

ROUSSEAU, J.-J. (1976) [1762] *The Social Contract*. London: Penguin.

ROUSSEAU, J.-J. (2004) [1755] *Discourse on the Origins of Inequality*. New York: Dover Publications.

RUGGIERO, G. DE (1966) *The History of European Liberalism*. Boston, MA: Beacon Press.

RUMMEL, J. D. (2015) Providing for sustainable exploration and use of outer space environments. In *Human Governance Beyond Earth: Implications for Freedom* (ed: Cockell, C. S.), pp. 139–47. Heidelberg: Springer.

RUSSELL, B. (1919) *Proposed Roads to Freedom*. New York: Henry Holt and Company.

RUSSELL, B. (2004) *Power*. New York: Routledge.

RUSSELL, B. (2010) *Freedom and Organisation*. London: Routledge.

SAGAN, C. (1973) Planetary engineering on Mars. *Icarus* 20, 513–14.

SALLUST. (2007) [various essays written between 86 and 35 BCE] *Catiline's War, The Jugurthine War, Histories*. London: Penguin.

SANDOZ, E. (1993) *The Roots of Liberty: Magna Carta, Ancient Constitution, and the Anglo-American Tradition of the Rule of Law*. Indianapolis, IN: Liberty Fund.

SCHMIDT, N., BOHACEK, P. (2021) First space colony: What political system could we expect? *Space Policy* 56, article no. 101426.

Schmidt, P., Newman, D., Hodgson, E. (2001) *Modeling Space Suit Mobility: Applications to Design and Operations*, SAE technical paper, 2001–01–2162.

Schmidtz, D. (1991) *The Limits of Government: An Essay on the Public Goods Argument*. Boulder, CO: Westview Press.

Schulte, P., Alegret, L., Arenillas, I., Arz, J. A., Barton, P. J., Bown, P. R., Bralower, T. J., Christeson, G. L., Claeys, P., Cockell, C. S., Collins, G. S., Deutsch, A., Goldin, T. J., Goto, K., Grajales-Nishimura, J. M., Grieve, R. A. F., Gulick, S. P. S., Johnson, K. R., Kiessling, W., Koeberl, C., Kring, D. A., MacLeod, K. G., Matsui, T., Melosh, J., Montanari, A., Morgan, J. V., Neal, C. R., Nichols, D. J., Norris, R. D., Pierazzo, E., Ravizza, G., Rebolledo-Vieyra, M., Reimold, W. U., Robin, E., Salge, T., Speijer, R. P., Sweet, A. R., Urrutia-Fucugauchi, J., Vajda, V., Whalen, M. T., Willumsen, P. S. (2010) The Chicxulub asteroid impact and mass extinction at the Cretaceous-Paleogene boundary. *Science* 327, 1214–18.

Schumpeter, J. A. (2008) *Capitalism, Socialism and Democracy*. New York: Harper.

Schwartz, J. S. J. (2015a) Fairness as a moral grounding for space policy. In *The Meaning of Liberty Beyond Earth* (ed: Cockell, C. S.), pp. 69–89. Heidelberg: Springer.

Schwartz, J. S. J. (2015b) Rendezvous with research: Government support of science in a space society. In *Human Governance Beyond Earth: Implications for Freedom* (ed: Cockell, C. S.), pp. 197–211. Heidelberg: Springer.

Schwartz, J. S. J. (2016) Lunar labor relations. In *Dissent, Revolution and Liberty Beyond Earth* (ed: Cockell, C. S.), pp. 41–58. Heidelberg: Springer.

Schwartz, J. S. J. (2018) Worldship ethics: Obligations to the crew. *Journal of the British Interplanetary Society* 71, 53–64.

Schwartz, J. S. J. (2020) *The Value of Science in Space Exploration*. Oxford: Oxford University Press.

Schwartz, J. S. J. (2021) A right to return to Earth? Emigration policy for the lunar state. In *The Human Factor in Settlement of the Moon* (eds: Rappaport, M. B., Szocik, K.), Heidelberg: Springer (in press).

SCHWIETERMAN, E.W., KIANG, N.Y., PARENTEAU, M.N., HARMAN, C.E., DASSARMA, S., FISHER, T.M., ARNEY, G.N., HARTNETT, H.E., REINHARD, C.T., OLSON, S.L., MEADOWS, V.S., COCKELL, C.S., WALKER, S.I., GRENFELL, J.L., HEGDE, S., RUGHEIMER, S., HU, R., LYONS, T.W. (2018) Exoplanet biosignatures: A review of remotely detectable signs of life. *Astrobiology* 18, 663–708.

SEN, A. (2001) *Development as Freedom*. Oxford: Oxford University Press.

SHUE, H. (2020) *Basic Rights*. Princeton, NJ: Princeton University Press.

SIDNEY, A. (1996) [1689] *Discourses Concerning Government*. Indianapolis, IN: Liberty Fund.

SIEDENTOP, L. (2014) *Inventing the Individual: The Origins of Western Liberalism*. London: Penguin Books.

SIMONTON, D. K. (2004) *Creativity in Science*. Cambridge: Cambridge University Press.

SINGER, P. (1973) *Democracy and Disobedience*. Oxford: Oxford University Press.

SKINNER, Q. (2004) *Liberty Before Liberalism*. Cambridge: Cambridge University Press.

SMITH, A. (1991) [1776] *An Inquiry into the Nature and Causes of the Wealth of Nations. Books I–IV*. New York: Everyman Library.

SMITH, A. (1999) [1776] *An Inquiry into the Nature and Causes of the Wealth of Nations. Books IV–V*. London: Penguin Books.

SMITH, A. (2010) [1759] *The Theory of Moral Sentiments*. London: Penguin Books.

SMITH, G. H. (2013) *The System of Liberty*. Cambridge: Cambridge University Press.

SMITH, K. C. (2016) Cultural evolution and the colonial imperative. In *Dissent, Revolution and Liberty Beyond Earth* (ed: Cockell, C.S.), pp. 169–187. Heidelberg: Springer.

SOLZHENITSYN, A. (1997) *The Gulag Archipelago*. Boulder, CO: Westview Press.

SPIRO, M. E. (1981) *Kibbutz: Venture in Utopia*. Cambridge, MA: Harvard University Press.

SPUDIS, P. D. (2016) *The Value of the Moon: How to Explore, Live, and Prosper in Space Using the Moon's Resources*. Washington, DC: Smithsonian Books.

SRIDHAR, K., VANIMAN, B. (1997) Oxygen production on Mars using solid oxide electrolysis. *Solid State Ion* 93, 321–8.

STASAVAGE, D. (2020) *The Decline and Rise of Democracy*. Princeton, NJ: Princeton University Press.

STEEL, T. (1965) *The Life and Death of St Kilda*. Edinburgh: National Trust of Scotland (R & R Clark Ltd).

STEVENS, A. (2015) The price of air. In *Human Governance Beyond Earth: Implications for Freedom* (ed: Cockell, C. S.), pp. 51–61. Heidelberg: Springer.

STEVENS, D. J., WARD, C. S. (1997) College education and recidivism: Educating criminals is meritorious. *Journal of Correctional Education* 48, 106–11.

SYCHEV, V., LEVINSKIKH, M. A., PODOLSKY, I. G. (2008) Biological component of life support systems for a crew in long-duration space expeditions. *Acta Astronautica* 63, 1119–125.

SZOCIK, K., BRADDOCK, M. (2019) Why human enhancement is necessary for successful human deep-space missions. *The New Bioethics* 25, 295–317.

SZOCIK, K., WÓJTOWICZ, T. (2019) Human enhancement in space missions: From moral controversy to technological duty. *Technology in Society* 59, 101156.

TACHIBANA, K. (2019) A Hobbesian qualm with space settlement. *Futures* 110, 28–30.

TALMON, J. L. (1961) *The Origins of Totalitarian Democracy*. New York: Frederick A. Praeger.

TAWNEY, R. H. (1975) [1922] *Religion and the Rise of Capitalism*. Harmondsworth, UK: Penguin Books.

TAYLOR, F. W. (2014) *The Principles of Scientific Management*. Greenwood, WI: Suzeteo Enterprises.

TEEGARDEN, D. A. (2014) *Death to Tyrants! Ancient Greek Democracy and the Struggle against Tyranny*. Princeton, NJ: Princeton University Press.

THUCYDIDES (1972) [early fourth century BCE] *History of the Peloponnesian War*. London: Penguin.

TÖNNIES, F. (2001) [1887] *Community and Civil Society*. Cambridge: Cambridge University Press.

TROUTMAN, P. A., BETHKE, K., STILLWAGEN, F., CALDWELL, D. L., MANVI, R., STRICKLAND, C., KRIZAN, S. A. (2003) *Revolutionary Concepts for Human Outer Planet Exploration (HOPE)*. Report from Space Technology and Applications International Forum, Albuquerque, NM.

TURFAN, B. (1995) *Emerging Democracies and Freedom of Information*. London: Library Association Publishing.

UHLIR, P. F., BISHOP, W. P. (1986) Wilderness and space. In *Beyond Spaceship Earth* (ed: Hargrove, E. C.), pp. 183–210. San Francisco: Sierra Club Books.

VACCA, J.S. (2004) Educated prisoners are less likely to return to prison. *Journal of Correctional Education* 55, 297–305.

VAN PARIJS, P. (1997) *Real Freedom for All: What (if anything) Can Justify Capitalism?* Oxford: Oxford University Press.

VIGNE, J. DE (2015a) Are we there yet? Pursuing a utopian ideal in education in space. In *Human Governance Beyond Earth: Implications for Freedom* (ed: Cockell, C. S.), pp. 225–33. Heidelberg: Springer.

VIGNE, J. DE (2015b) Education and liberty in space. In *The Meaning of Liberty Beyond Earth* (ed: Cockell, C. S.), pp. 219–26. Heidelberg: Springer.

VIGNE, J. DE (2016) Sturm und Drang: Guaranteeing teenage liberty in space: An exploration of the joys and dangers of adolescence in an off world context. In *Dissent, Revolution and Liberty Beyond Earth* (ed: Cockell, C. S.), pp. 143–54. Heidelberg: Springer.

VILE, M. J. C. (1998) *Constitutionalism and the Separation of Powers*. Indianapolis, IN: Liberty Fund.

WAMELINK, G. W. W., FRISSEL, J. Y., KRIJNEN, W. H. J., VERWOERT, M. R., GOEDHART, P. W. (2014) Can plants grow on Mars and the Moon: A growth experiment on Mars and Moon soil simulants. *PLoS ONE* 9, article no. e103138.

WATTS, E. J. (2018) *Mortal Republic: How Rome Fell into Tyranny*. New York: Basic Books.

WEBER, M. (2007) [1905] *The Protestant Ethic and the Spirit of Capitalism*. London: Routledge.

WELLS-JENSEN, S., MIELE, J. A., BOHNEY, B. (2019) An alternate vision for colonization. *Futures* 110, 50–3.

WENER, R. E. (2014) *The Environmental Psychology of Prisons and Jails: Creating Humane Spaces In Secure Settings.* Cambridge: Cambridge University Press.

WHITE, F. (1987) *The Overview Effect: Space Exploration and Human Evolution.* New York: Houghton-Mifflin.

WHITMAN COBB, W.N. (2021) *Privatizing Space.* New York: Routledge.

WILLIAM OF OCKHAM (2001) [early fourteenth century] *A Short Discourse on Tyrannical Government.* Cambridge: Cambridge University Press.

WILLIAMSON, M. (2006) *Space: The Fragile Frontier.* Reston, VA: American Institute of Aeronautics and Astronautics.

WILSON, B. (2009) *What Price Liberty? How Freedom was Won and is Being Lost.* London: Faber and Faber.

WOLLSTONECRAFT, M. (2001) [1792] *The Vindications.* Ontario: Broadview Literary Texts.

YOUNG, F. (1988) Labor relations in space: An essay in extraterrestrial business ethics. *The Monist* 71, 114–29.

ZIMBARDO, P. (2008) *The Lucifer Effect: How Good People Turn Evil.* London: Rider.

ZINN, H. (2013) *Disobedience and Democracy: Nine Fallacies on Law and Order.* Chicago, IL: Haymarket Books.

ZUBRIN, R. (2011) *The Case for Mars: The Plan to Settle the Red Planet and Why We Must.* New York: Free Press.

INDEX